SEAFOOD BIOCHEMISTRY

SEAFOOD BIOCHEMISTRY

Edited by

Dr. Sandeep Gore and
Dr. Subal Kumar Ghosh

Delve
PUBLISHING

www.delvepublishing.com

Seafood Biochemistry

Dr. Sandeep Gore and Dr. Subal Kumar Ghosh

Delve Publishing
4164 Lakeshore Road
Burlington, ON L7L 1A4
Canada
www.delvepublishing.com
Email: orders@arclereducation.com

ISBN: 978-1-77956-372-9 (Hardcover)

ABOUT THE EDITORS

Dr. Sandeep B. Gore (1979) is presently serving as Assistant professor in the Department of Fisheries Engineering, College of Fisheries Science, Nagpur. He has obtained his Bachelor of Fisheries Science (B.F.Sc.) in 2002 from Dr. Balasaheb Sawant Konkan Krishi Vidyapeeth, Dapoli. He has done his Masters of Fisheries Science (M.F.Sc.) in Fish Processing Technology in 2004 from the Dr. Balasaheb Sawant Konkan Krishi Vidyapeeth, Dapoli and Ph. D. in Fish and Fisheries Science with specialization in Post-Harvest Technology from ICAR-CIFE, Mumbai in 2022. He started his career as Project Assistant from 2005-2007 in Fishery Survey of India, Govt. of India. Since 2007, he has been working as Assistant professor in the Department of Fisheries Engineering, College of Fisheries Science, Nagpur under the Maharashtra Animal and Fishery Science University, Nagpur. He has published more than 30 research papers in scientific journals of National and International repute. He has participated in many national and international seminars and conferences.

Dr. Subal Kumar Ghosh (1989) is presently serving as an Assistant Fisheries Officer, at the Fisheries & Animal Resources Development Department Govt. of Odisha. He has obtained a Bachelor of Fisheries Science (B.F.Sc.) in 2015 from W.B.U.A.F.S, Kolkata. He completed his Master's in Fisheries Science (M.F.Sc.) in 2017 from the Central Institute of Fisheries Education (CIFE), Mumbai, and his Ph. D. in Fisheries Post-Harvest Technology from CIFE, Mumbai in the year 2023. He started his career as an Assistant Professor in the Dept. of Fish Processing Technology at the School of Fisheries, CUTM, Paralakhemundi, Odisha, in the year 2021. Since 2023, he has been working as an Assistant Fisheries officer at FARD Dept. Govt. of Odisha. He has expertise in the development of fish-based value-added products and by-products, as well as physicochemical and microbiological techniques for fish and fisheries products. Moreover, he has also been exposed to special field training and participated in many national and international seminars, and skill development learning programs in the field of fisheries science. He has more than 13 years of experience in the field of fisheries science and published more than 20 research articles in national and international journals of repute.

TABLE OF CONTENTS

LIST OF FIGURES

ABSTRACT

Seafood Biochemistry is a comprehensive book that explores the intricate biochemical processes in seafood products, essential for ensuring the safety, quality, and nutritional value of seafood. The book covers topics such as detecting contaminants, determining freshness, and understanding the factors that influence seafood quality. It is targeted towards researchers, scientists, and professionals in food science, nutrition, and fisheries. By studying seafood biochemistry, readers can enhance the safety, quality, and nutritional value of seafood, benefiting both consumers and producers in the industry. Overall, the book serves as a valuable resource for understanding the biochemistry of seafood.

PREFACE

The field of seafood biochemistry is incredibly intriguing as it delves into the intricate chemical processes and compounds present in different varieties of sea animals. Aquatic creatures, such as fish and shellfish, possess a multitude of complex molecules that are vital for their growth, development, and overall well-being.

Seafood Biochemistry provides a thorough examination of the biochemical composition and properties of different seafood products. This text explores the complex chemical processes that take place in seafood, providing insights into the nutritional value, flavor development, and quality of these products. It effectively connects the scientific community with the seafood industry, enabling researchers, food technologists, and industry professionals to make well-informed choices regarding seafood production, processing, and preservation. Additionally, a thorough understanding of seafood biochemistry plays a vital role in ensuring the successful implementation of food safety and quality control measures.

The information contained in *Seafood Biochemistry* is of utmost importance, given the increasing global demand for seafood. As demand rises, it becomes imperative to streamline production methods and enhance the nutritional value of seafood products. This book offers a comprehensive analysis of seafood's biochemical composition, providing valuable insights that can fuel innovation and promote sustainability in the industry.

Seafood Biochemistry stands out for its distinct emphasis on the biochemical aspects of seafood. It provides a comprehensive analysis of the molecular components that contribute to seafood's sensory attributes and health benefits. This book offers a multidisciplinary approach to seafood biochemistry, exploring the intricate relationships among proteins, lipids, carbohydrates, vitamins, and minerals.

This book will greatly interest researchers in food science, nutrition, and biochemistry, as well as professionals in the seafood industry, including seafood processors, quality control managers, and product developers. Students of food science, marine biology, or nutrition will also find valuable knowledge and insights to enhance their understanding of the subject matter.

Through this book, readers will develop a deeper understanding of the biochemical processes that influence the quality and safety of seafood products. They will gain insights into the significance of proteins in texture formation, the impact of lipid oxidation on flavor development, and the importance of vitamins and minerals in nutritional value. With this knowledge, readers will have a stronger foundation to address challenges in seafood production, processing, and preservation, ultimately leading to higher product quality and increased consumer satisfaction.

Ultimately, *Seafood Biochemistry* proves to be an invaluable resource, providing an in-depth examination of the biochemical composition and properties of seafood. This book serves as a bridge between the scientific community and the seafood industry, offering essential knowledge and fostering a deeper understanding of the subject matter. Highly recommended for individuals who wish to expand their knowledge of seafood products at the molecular level, its specialized focus on seafood biochemistry sets it apart from other publications in the field.

Chapter 1

Introduction to Seafood Biochemistry

Contents

INTRODUCTION

Seafood, central to our diets, traditions, and economies, represents one of the most important elements of food from the vastness of the oceans. Seafood hosts diverse biochemical processes that give it its unique taste, texture, and high nutritional value. These are the hidden processes that give seafood so much of its character and make it truly beneficial to our health.

Imagine the subtle flavors of any fish, brought out by the delicate balance between proteins and fats. Or think about how some compounds work to preserve freshness and fight off spoiling. These are not just coincidences of nature, but complex biochemical reactions. Such reactions are important in that they ensure seafood remains a healthy and safe food supply from the ocean to our table (Figure 1.1) (Hui et al., 2012).

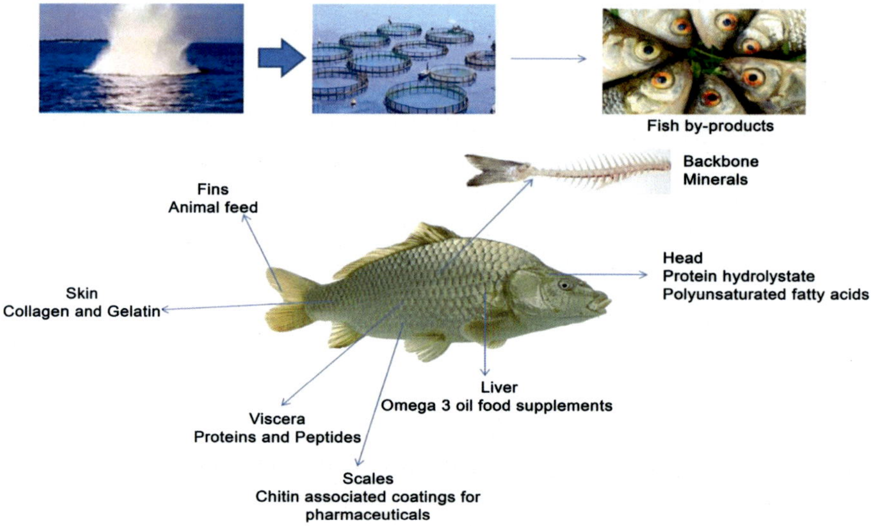

Figure 1.1. *Fish and its by-products as nutraceuticals.*

Source: Nag M, Lahiri D, Dey A, Sarkar T, Pati S, Joshi S, Bunawan H, Mohammed A, Edinur HA, Ghosh S, and Ray RR (2022) Seafood Discards: A Potent Source of Enzymes and Biomacromolecules With Nutritional and Nutraceutical Significance. Front. Nutr. 9:879929. doi: 10.3389/fnut.2022.879929.

This book on seafood biochemistry is not only for science but also for appreciating these wonders of nature that make seafood special. Through such studies, we would understand how to uphold quality, enhance health benefits, and ensure sustainability. We take a journey into the realm of seafood biochemistry to explore some of those very interesting ways through which life in the ocean impacts our daily lives and our future.

1.1. DEFINITION OF SEAFOOD BIOCHEMISTRY

Seafood biochemistry is a specialized area of biochemistry aimed at explaining chemical procedures and materials that take place in seafood species, including fish, crustaceans, mollusks, and other marine organisms. Studies that fall under this field range from the molecular structure of components found in seafood to biochemical reactions that affect its quality and flavor, and thus its nutritional value. Investigations of these processes allow inference of ways to maintain and improve the safety, quality, and nutritional benefits of seafood. We shall later differentiate it from the general concept of biochemistry.

1.1.1. Fundamental Concepts and Terminologies

Basically, seafood biochemistry studies different kinds of biomolecules like proteins, lipids, carbohydrates, vitamins, and minerals in seafood. Each of them plays a vital role in defining the nutritional profile and the sensory features of seafood include:

- **Proteins:** Being one of the major constituents of seafood, proteins help in providing not only a significant supply of nutrition but also the texture and flavor. In the case of seafood, the most common proteins are the myofibrillar proteins involved in the structure of muscles and the metabolic sarcoplasmic proteins. Specialized proteins, the enzymes, have a vital role in causing post-mortem changes, which relate to the freshness or shelf life of seafood.
- **Lipids:** Seafood is the major source of health-giving lipids, particularly omega-3 fatty acids such as EPA, eicosapentaenoic acid, and DHA, docosahexaenoic acid. These types of fats play a crucial role in human health; they function in the brain and lessen the risk of chronic diseases. The nature of the lipid profile varies between species, showing a high degree of difference and being influenced by diet and environment.
- **Carbohydrates:** While they are less abundant as compared to proteins and lipids, seafood carbohydrates are of immense importance as storage forms of energy and can have far-reaching impacts on the texture and flavor of flesh. The major carbohydrate in seafood is glycogen, which is associated with biochemical changes that take place upon death of fish.
- **Vitamins and Minerals:** Seafood forms an excellent source of fat-soluble and water-soluble vitamins and some important minerals, such as vitamin D and B12, selenium, and iodine. These micronutrients have key functions in physiological functioning and

help to explain some of the health benefits attributed to seafood consumption.

- **Water:** A high proportion of seafood is composed of water. The amount ranges from 60% in some to more than 80% in others. Water content contributes to texture, juiciness, and biochemical reactions after catching/harvest.

1.1.2. Differences from General Biochemistry

While general biochemistry covers a very broad range of topics in terms of the chemical processes taking place in all living organisms, seafood biochemistry deals only with those biochemical properties and processes related to the specificity of marine life. A number of peculiarities distinguish seafood biochemistry from general biochemistry.

Seafood forms a unique biochemical composition in its lipid and protein content. Characteristic and distinguishing seafood from other sources of food is the existence of long-chain polyunsaturated fatty acids, specific amino acids, and unique enzymes. This shall be very important in appreciating the special health benefits and culinary qualities of seafood (Ochiai and Ozawa, 2020).

Temperature of water, salinity, and diet are among the major environmental factors that exert a great influence on the biochemical properties of seafood. These factors modify the metabolic processes of aquatic organisms, causing changes in their nutrient composition, flavor, and texture. This is different from terrestrial animals, whose diet and habitat conditions are more or less standardized.

Changes in seafood at the biochemical level are drastic after harvesting. This includes enzymatic degradation, lipid oxidation, and microbial growth. These changes are important in explaining the seafood problems related to spoilage, preservation, and safety. Seafood is a highly perishable food compared with many terrestrial foods, and its biochemistry directly affects the shelf life and quality of it.

Seafood's taste, texture, and aroma—the sensory attributes are intrinsically related to its biochemical composition. For example, seafood's umami taste is mainly due to free amino acids and nucleotides. These attributes vary greatly depending on the type of seafood product and are determined by intrinsic factors such as species and age and extrinsic factors like the type of storage.

- **Health Implications:** These seafood biochemistries also pertain to the health implications of their consumption, especially concerning the intake of their beneficial nutrients like omega-3 fatty acids and possible risks due to contaminants such as heavy metals or biotoxins.

This aspect is very important in view of the growing interest of consumers in health matters and wellness.

1.2. IMPORTANCE OF STUDYING SEAFOOD BIOCHEMISTRY

Seafood Biochemistry is an important field of study bridging the gap among marine biology, nutrition, food science, and public health. It provides critical insight into the chemical processes governing seafood quality, safety, and nutritional value. Biochemical knowledge about seafood is very important for many reasons, ranging from ensuring the safety of food to improving food quality and promoting public health through better nutrition (Undeland et al., 2009).

Seafood biochemistry has a close linkage with food safety and quality. Seafood is a very perishable produce that undergoes lots of biochemical changes, thus affecting its safety and quality post-harvest. The biochemical composition in seafood including proteins, lipids, carbohydrates, vitamins, minerals, and water all have significant changes by enzymatic reactions and microbial growth, along with environmental factors. These changes may then lead to spoilage, the development of off-flavors, and the formation of harmful compounds such as biogenic amines and toxins. Having understood these processes, food scientists and technologists are at a vantage position to devise ways of preventing or minimizing spoilage and contamination, hence ensuring that seafood remains safe for human consumption.

For example, unwanted compounds may be formed during the enzymatic degradation of proteins and lipids in fish, influencing sensory attributes and shelf-life quality. Knowing these biochemical reactions allows the development of methods of preservation for example freezing, salting, smoking and additives, to help maintain the quality and extend the shelf life of seafood products. Furthermore, seafood biochemistry gains importance in the detection and quantification of contaminants, heavy metals, persistent organic pollutants, and microbial pathogens. The biochemical interaction between the aforementioned contaminants with seafood components needs to be understood to develop adequate programs for monitoring and control to assure the protection of public health.

Other relevant topics of seafood biochemistry are those dealing with nutrition and public health. Seafood is one of the better sources of high biological value proteins, essential fatty acids, vitamins, and minerals and forms, therefore, an integral part of a balanced diet. Studies on seafood biochemistry will help in detailing the nutritional features of different species of seafood so that proper dietary recommendations and policies on public health can be made. For example, omega-3 fatty acids, particularly EPA and DHA, exist in large amounts in many seafood types. They are known to have anti-inflammatory effects and benefits for cardiovascular health. Research on the biochemical pathways leading to their accumulation in seafood can thus realize which species best concentrate these fatty acids and therefore advise them as part of a healthy diet.

The biochemistry of seafoods further explains bioavailability is vital in understanding the degree that which the body absorbs and utilizes a nutrient from the amount consumed through seafood. Influences would thus emanate from the chemical form of a nutrient or due to the presence of other dietary components, while another individual factor would be the physiological state. For instance, while the iron content might be similar in seafood and plant-based foods, its bioavailability from the former is often much higher and, therefore, very essential in populations at risk of iron deficiency. The ability to take account of such subtleties enables the production of more accurate dietary recommendations and makes it easier to avoid nutritional deficiencies in specific populations.

Besides the nutritional value of seafood, the biochemical composition of seafood also contributes to understanding and reducing the risks involved in seafood consumption. While seafood is a very nutritious food, it is also a source of some harmful substances, such as mercury, dioxins, and PCBs, which may become significant health risks, particularly for sensitive populations such as pregnant women and young children. By investigating the biochemical mechanisms for the accumulation of these contaminants in seafood, scientists set the guidelines for safe levels of consumption and determine practices for reducing exposure. For instance, knowledge of the trophic transfer of mercury in aquatic food chains has led to advisories regarding minimum consumption levels of some high-mercury species to avert risk to public health.

Seafood biochemistry is also instrumental in the development of functional foods and nutraceuticals that confer health benefits beyond that of basic nutrition. Bioactive peptides, phospholipids, and other seafood antioxidants display various potential health-promoting properties through anti-inflammatory, antihypertensive, and antioxidant activities. These compounds, once isolated and characterized, will add to further research in the development of new food products targeted toward special dietary or health needs, such as lowering chronic disease risk or retaining cognitive function with aging. The value that this enhances includes seafood products and a growing market of functional foods and nutraceuticals.

Seafood biochemistry has implications for the sustainability and conservation of marine resources. With the demand for seafood, there is an increased need to manage fisheries and aquaculture systems sustainably. Knowing the biochemical composition and nutritional value of various species of seafood will help in the selection for aquaculture, specifically those species that deliver high nutritional benefits but have lower environmental impacts. The investigation into seafood biochemical processes of spoilage may further open up new avenues for reducing food wastage, thereby contributing to sustainability.

1.3. HISTORICAL BACKGROUND OF SEAFOOD CHEMISTRY

History in the field of seafood biochemistry is anchored by our exploration as humanity to the diverse marine life. This field has achieved significant growth with huge trends over the past years. This development occurred parallel to advances in technology, better knowledge of marine ecosystems, and appreciation for the nutritional value of seafood. Throughout the journey of seafood biochemistry, a few historical milestones and contributions were laid down by influential players who played a very key role in shaping the field the way we know it today.

One of the early scientific studies about marine life in the 20th century concentrated mainly on taxonomy and ecology, giving little emphasis on the chemical composition of marine organisms. However, with the increasing global demand for seafood, especially after the world wars, came the need to determine the factors that influence the quality and safety of seafood. It was the beginning of the discipline of seafood biochemistry when scientists began studying the chemical makeup of seafood mainly proteins and lipids, which proved very important for human health.

Among the significant recent milestones of seafood biochemistry, one was the identification and characterization of omega-3 fatty acids in fish oils. Researchers like Dr. Hugh Sinclair and Dr. John Dyerberg spearheaded the study of these fatty acids in the prevention of cardiovascular diseases during the 1930s and 1940s (Truswell, 2010). These researchers helped in building a base for an understanding of the peculiar nutritional values of seafood that set it apart from other protein sources. This discovery became not only the stimulus of further studies on health benefits from seafood but also influenced dietary guidelines all around the world, having been put forward in favor of fish consumption being heart-healthy.

In the middle of the 20th century, a good number of analytical techniques were developed for the more accurate quantification and characterization of seafood biochemical components. Chromatography and mass spectrometry allowed for the identification and determination of concentrations for proteins, lipids, carbohydrates, and other bioactive compounds in marine organisms. In fact, these technological developments were incumbent upon the progress of this field by providing tools necessary for investigating complex seafood biochemistry at the molecular level.

Influential contributors to the literature at that time, such as Dr. Harold Barnes and Dr. John Emmett, made very vital contributions toward laying the principles of seafood spoilage and preservation. The work done on enzymatic degradation and lipid oxidation in fish by these above mentioned people was of considerable value in influencing the shelf life and quality of seafood. This was of special importance to the seafood industry, since it brought about new and better storage and processing methods, thereby reducing losses and assuring seafood safety. In these two decades, the 1960s and 1970s, research in seafood biochemistry focused particularly on the investigation of contaminants and toxins present in seafood. The incident of Minamata disease in Japan, caused by mercury poisoning through the consumption of contaminated fish, brought realization for monitoring and insight into the accumulation of these substances of concern in marine food chains (Hachiya, 2006). Scientists including Dr. Masazumi Harada played an important role in the documentation of health effects from mercury exposure, which led to stricter regulations and more comprehensive research on environmental contaminants in seafood. It marked the beginning of further evolution toward holistic seafood biochemistry, in which not only the beneficial but also the harmful components of seafood were considered for human consumption.

The late 20th and early 21st centuries belonged to further diversification of the field with an enhanced interest in the functional properties of compartment-derived seafood compounds. Research began to examine the potential of bioactive peptides, antioxidants, and other compounds for health promotion and disease prevention. It also saw the beginning of aquaculture, bringing new challenges and opportunities in seafood biochemistry. People like Dr. Ronald Hardy provided our knowledge of fish nutrition and feed formulations that would ensure farmed seafood of high quality with optimum nutritional profiles.

The continued development of the field incorporates newer technologies like genomics, proteomics, and metabolomics, among others, which are propelling new discoveries in seafood biochemistry. These "omics" approaches give a more complete picture of the biochemical pathways in marine organisms and thus explain better their nutritional and health-related value. Today, through the exploration and research provided by scientists, such as Dr. Fereidoon Shahidi and Dr. David Heber, this field is being stretched to explore seafood's potential as a source of nutraceuticals and functional foods.

1.4. SCOPE AND SIGNIFICANCE OF SEAFOOD BIOCHEMISTRY

Seafood biochemistry represents an area of great importance in several industries that have benefited from a greater understanding of the biochemical processes intrinsic to marine organisms. The most direct application would probably be in the seafood industry itself, where knowing about biochemistry can be critical for the quality and safety of seafood products. This ranges from enzymatic reactions consequent to harvesting, which may result in spoilage, to advanced techniques of preservation that elongate shelf life without affecting its nutritional value. For instance, the application of natural antioxidants derived from seafood itself has become a valuable strategy in preventing lipid oxidation resulting in rancidity and off-flavors.

The biggest payoff in seafood biochemistry in terms of nutrition and health is in the development of dietary supplements and functional foods. Various bioactive compounds discovered by seafood biochemistry include omega-3 fatty acids, peptides, and marine-derived antioxidants, which have driven the preparation of supplements targeted at improving cardiovascular health, reducing inflammation, and even enhancing cognitive function (Soccol and Oetterer, 2003). These have been exploited by the nutraceutical industry, which markets them as part of a good diet to achieve optimum health outcomes. Knowledge of the nutrient profile of different species of seafood can also come in handy in the development of fortified food products that are meant to make up for some deficiencies in the diet of a particular population.

Other areas that have so far leveraged much from seafood biochemistry include pharmaceuticals and cosmetics. Marine organisms, therefore, turn out to be a rich source of bioactive compounds with potential therapeutic properties. For instance, some peptides and enzymes isolated from fish and shellfish revealed their potency in developing new drugs against hypertension and arthritis. Likewise, the cosmetic industry has tapped into the marine-derived ingredients in skin-protective and anti-aging properties from the sea, exploiting an oceanic wealth of bioactive compounds that are extractable from seaweeds, fish collagen, and other marine sources.

Extensive use of concepts in seafood biochemistry is widely applied in explaining fish nutrition, and improving efficiency in fish farming within aquaculture. Knowing the biochemical requirements of the different species allows one to formulate feeds in a way that will support fast growth without compromising the nutritional quality of farmed seafood. This becomes critical in ensuring that levels of certain beneficial compounds, like omega-3 fatty acids, are at least as high in farmed fish as in wild relatives. Furthermore, the comprehension of fish biochemical responses to different environmental stressors could be utilized to realize improved aquaculture practices for the better welfare of farmed species.

1.4.1. Interdisciplinary Connections

There are a number of fields and disciplines that directly relate to seafood biochemistry.

- Marine biology provides the core information on the organisms being discussed in seafood biochemistry. This entails knowledge of species' life cycles, habitats, and behaviors in the ocean, all pertinent to biochemical makeup. For instance, the type of food or habitat a fish species ingests will greatly affect its fatty acid profile and consequently its nutritional value. Hence, any study that combines both biological and chemical aspects of seafood must be undertaken collaboratively by marine biologists and biochemists.
- Food science pertains to seafood biochemistry in the development of new processing and preservation technologies. Food scientists use biochemical principles to help optimize these processes so that seafood continues to offer nutrient value as well as desirable sensory characteristics. For example, knowing how enzymes impact the development of texture and flavor in seafood products could help in the control of these factors in processing better. This interdisciplinary approach ensures the best quality in food products reaching consumers.
- Moreover, environmental science is relevant and gives insight into the biological changes of seafood due to environmental change. Water temperature, pollution, and other factors leading to ocean acidification can alter the biochemical makeup of organisms from the sea, rendering them unsuitable for consumption or low in nutritional value. In this manner, seafood biochemists, in association with environmental scientists, are better positioned to foresee and lighten the impacts of these changes on seafood supplies.
- Another area with strong interdisciplinary links in seafood biochemistry is public health. Development of dietary recommendations for public health policy needs to take into account knowledge about the biochemical properties and information about their biochemical influence on human health. Not only is such collaboration critical to developing these recommendations based on the best possible available scientific evidence, but it will also help to avoid nutritional deficiencies and decrease the risk of diet-related diseases.

1.5. METHODS USED IN STUDYING SEAFOOD BIOCHEMISTRY

Studying seafood biochemistry relies on several analytical techniques, instruments, and methodologies in the investigation of the complex chemical makeup of marine organisms and the biochemical processes that affect them. Such methods are imperative in assessing the nutritional content, quality, and safety, and the functional properties of seafood. Some of the most important analytical techniques and instruments include:

- **Chromatography:** This is more specifically gas chromatography and high-performance liquid chromatography, is among the most prominent techniques in seafood biochemistry. Gas chromatography (GC) is particularly applied to the separation and analysis of volatile compounds such as fatty acids and other lipids. For example, in determining the fatty acid pattern of various species of seafood, which is quite useful in nutritional studies. On the other hand, high-pressure liquid chromatography (HPLC) is used for the separation and quantification of non-volatile compounds, including amino acids, peptides, and vitamins. Such techniques permit very accurate analysis of complex mixtures and deliver very valuable information on the biochemical composition of seafood.
- **Mass Spectrometry (MS):** It is usually combined with chromatography and has very strong prepotency in the identification and quantification of compounds at very low concentrations. In seafood biochemistry, MS is used for the detection of contaminants such as pesticides and heavy metals and also for the detection of bioactive compounds like marine-derived peptides and antioxidants. The ability of MS to provide molecular weight and structural information makes it an indispensable tool while characterizing the myriad substances present in seafood.
- **Spectroscopic Techniques:** They may be used to identify the structural and chemical features of the compound. Such techniques include nuclear magnetic resonance (NMR) spectroscopy and infrared (IR) spectroscopy. NMR spectroscopy can give information about the structure of a complex molecule like proteins and lipids and, therefore, deliver information related to their functional properties. IR spectroscopy may determine functional groups and hence describe the molecular composition of seafood, which helps in the evaluation process for freshness and quality.
- **Enzyme-Linked Immunosorbent Assay (ELISA):** It is a technique

developed for the identification of a certain protein and other biomolecules (Shibahara, Uesaka, Wang, Yamada, and Shiomi, 2013). It comes in handy in the detection of allergens, where specific antibodies are used against proteins responsible for eliciting an allergic response. This technique ensures that seafood products are safe for consumers, especially those with food allergies.

- **Polymerase Chain Reaction, PCR, and Sequencing:** These techniques are applied to the study of seafood species' genetic material. Among others, these are key techniques in species identification for labeling and regulatory purposes and in the detection of possible pathogens that might cause health risk. Sequencing techniques can be applied in expressing genes involved in nutritional and stress responses in aquaculture species.

1.5.1. Experimental Design and Data Analysis

Experimental design and data analysis are the two important units in seafood biochemical research. They help to ensure studies are scientifically sound and results are reliable and reproducible.

One of the prerequisites to producing meaningful and interpretable data is a well-planned experimental design. This includes the selection of the right controls, randomization of samples against bias, and replication to account for variability. Experimental designs in seafood biochemistry mostly involve variables of temperature, storage conditions, and processing methods so that the effect of specific biochemical factors can be isolated. For example, studies of the effects of freezing on fish lipids must ensure that methods and periods for thawing are carefully controlled to ensure accurate results.

Proper sampling is important to derive a representative data set. In seafood biochemistry, it may involve sampling at different levels of the organism or during different stages of processing. Studies related to environmental contaminants also require sampling from different geographic locations or water conditions. The sampling techniques chosen in this regard could, therefore, to a large extent affect the accuracy and relevance of the findings.

Most of the time, biochemical data analysis will go hand in hand with statistical methods to interpret the results and their significance. Examples include analysis of variance, regression analysis, and multivariate analysis. In seafood biochemistry, multivariate analysis can be very efficient while studying the interrelation between various biochemical variables; for example, the correlation of lipid oxidation with sensory attributes like flavor and texture. Because of the growing complexity and data volume of seafood biochemistry, the use of tools in bioinformatics and computational science is increasingly important. These tools will allow for the analysis and visualization of large datasets obtained from studies in genomics and proteomics. The functionality of proteins and peptides can be predicted with such bioinformatics approaches from their sequences alone, relevant in searching for new bioactive compounds.

1.6. CURRENT TRENDS IN SEAFOOD BIOCHEMISTRY RESEARCH

Seafood biochemistry is a dynamic field where researchers are constantly pushing into new frontiers that can help improve our understanding of the biochemical processes controlling seafood quality, safety, and nutritional value. Currently, one of the most prominent emerging areas of study is the exploration of bioactive compounds in seafood. Thus, researchers come up with the identification and characterization of peptides, omega-3 fatty acids, and other bioactive molecules that, when taken, can have immense health benefits beyond simple nutrition. Research is underway to study these molecules for probable roles they may play in the prevention of major chronic diseases such as heart disease, cancer, or neurodegenerative disorders. This is further supported by an enhanced interest in functional foods and nutraceuticals that, besides nutritional value, should provide particular human health benefits.

The second major trend in seafood biochemistry research is related to studying the influence of environmental change on seafood quality and safety. Climate change, pollution, and other environmental factors have been described as influencing the biochemical composition of marine organisms. For instance, sea temperature rise can result in changes to the fatty acid profiles of fish, and pollution may promote the bioaccumulation of some dangerous contaminants like heavy metals and persistent organic pollutants. Research into these effects provides insight into the aspects that affect the nutritional value of seafood and their safety for consumption, with potential mitigation strategies.

Another major trend is that seafood biochemistry is becoming associated with omics technologies, including genomics, proteomics, and metabolomics. Such advanced techniques enable the global analysis of the genetic, protein, and metabolic profiles of marine organisms, providing insights into their biochemical responses to environmental, dietary, and other stressors. For example, using genomics, genes responsible for the synthesis of nutritional components like omega-3 fatty acids can be identified, while with proteomics, protein expression patterns related to different seafood quality traits can be revealed. The study of small metabolites resulting from metabolic processes, known as metabolomics, can give an exact snapshot of the biochemical state of seafood and thus help in quality control and safety assessments (Figure 1.2).

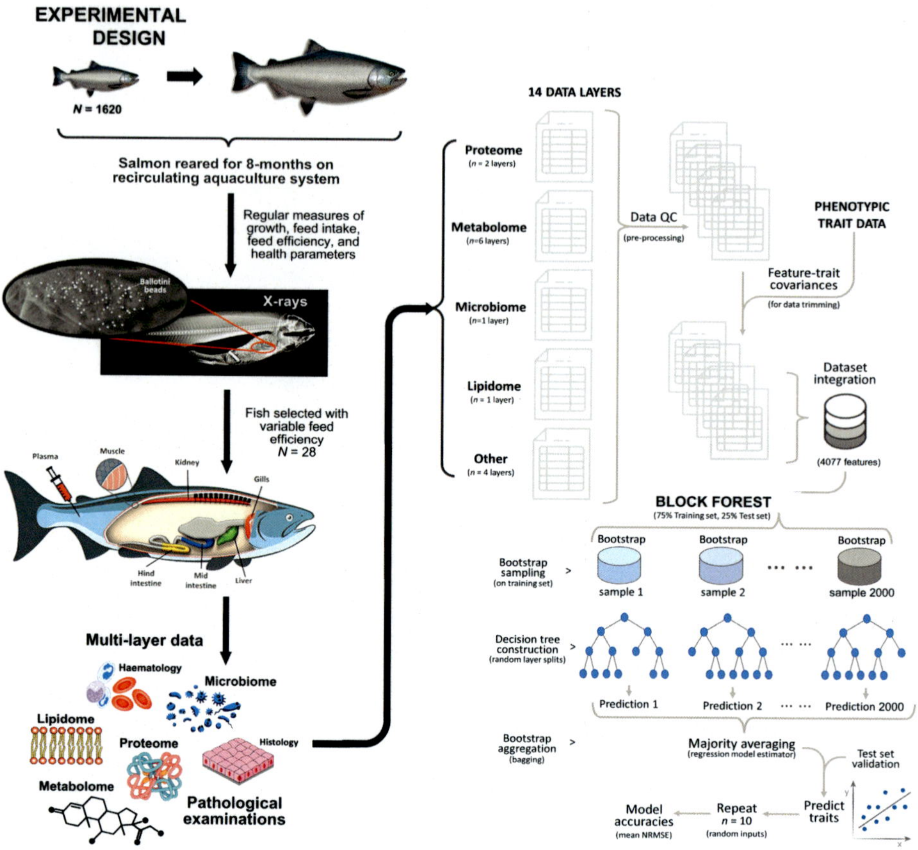

Figure 1.2. *Omics approaches to study fish.*

Source: Young T, Laroche O, Walker SP, Miller MR, Casanovas P, Steiner K, Esmaeili N, Zhao R, Bowman JP, Wilson R, et al. Prediction of Feed Efficiency and Performance-Based Traits in Fish via Integration of Multiple Omics and Clinical Covariates. Biology. *2023; 12(8):1135. https://doi.org/10.3390/biology12081135.*

Recent technological advances also fuel innovation in seafood biochemistry research. For example, high-throughput sequencing technologies have transformed studies on the microbiome community associated with seafood. Specifically, it deals with the microbiota of seafood, which may have an impact on quality and safety. It enables research into the microbiome to improve methods for predicting and controlling spoilage and may point out the pathogens that are present. Mass spectrometry and chromatography have also improved the identification and quantification of bioactive compounds and contaminants, allowing for more complete and accurate analysis. Moreover, there is increased interest in new approaches for seafood processing and preservation using sustainable and green methods. Scientists are now developing natural preservatives from plants and marine organisms, which could prolong the shelf life of seafood while keeping its nutritional value. This trend, in particular, is impelled as more customers yearn for clean label products—those free of synthetic additives and chemicals. Research in the development of biodegradable packaging materials often combined with antimicrobial agents is going on and will continue to play an important role in seafood packaging to reduce the impact of seafood packaging on the environment without compromising the safety of the product.

Biotechnology applied to seafood biochemistry opens further prospects for research and development in this area. Genetic engineering methods are being studied to improve the nutritional value of farmed fish, enhance levels of beneficial fatty acids, or reduce the accumulation of toxic elements like mercury. Biotechnology applications play a key role in the development of new products based on seafood, such as cultured fish, which promise sustainable alternatives to traditional fishing and aquaculture.

1.7. CHALLENGES IN STUDYING SEAFOOD BIOCHEMISTRY

1.7.1. Environmental and Biological Variability

Environmental variability influences the biochemical composition of marine organisms. Sea temperature, salinity, pH, and pollution in fact may vary significantly between different marine environments and may change seasonally in the same region. All this environmental variability may give rise to differences in the composition of nutrients and contaminants in seafood, even within the same species. For example, fish caught in colder waters generally contain more omega-3 fatty acids than fish from warmer waters. Similarly, contaminants such as heavy metals bioaccumulate variably depending on the water pollution levels where seafood was sourced.

Biological variability further complicates the study of seafood biochemistry. This would mean genetic diversity within species, age and size differences, and variability in diet and metabolism. For example, the nutrient contents of wild-caught fish are likely to be significantly different from those of farmed fish, mainly as a result of diet and living conditions. Even within farmed populations, variables like feed composition, stocking density, and water quality might have large effects on the biochemical properties of the fish. The biological diversity, thus poses a challenge to the standardization of research findings and universality of guidelines within the seafood industry.

Another intrinsic factor contributing to variation in biochemical composition is processing and handling variability after harvesting. Processing methods such as freezing, drying, smoking, or canning and the conditions under which these are conducted will influence the nutritional value and safety of seafood. For example, lipid oxidation takes place during freezing and storage and can affect the flavor of fish and its nutritional quality. The intrinsic variation in the techniques used for processing makes it very hard to compare studies and establish homogeneous quality standards.

1.7.2. Limitations of Current Methodologies

Advanced analytical techniques still feature issues such as sensitivity, specificity, precision and accuracy. The detection of certain low-contaminated levels, for example, heavy metals or pesticides, requires highly sensitive instruments that may always not be available in every research setting. In addition, the complexity of the matrices of seafood itself might further interfere with the accuracy of the analytical method, leading to an underestimate or overestimate of nutrient levels or contaminants.

Another of the methodological challenges is the standardization across different studies for techniques and protocols. Lack of standardized methodology may be the reason for a certain discrepancy in data and inferred conclusions, again, making it difficult to compare results from two different research groups. This is particularly relevant in studies of new, emerging contaminants or bioactive compounds for which methodologies of detection and quantification are under continuous development. This thus brings out the variability in analytical methods, sometimes problematic for regulatory assessment and setting guidelines for safe seafood consumption.

Moreover, biochemical data is often based on the multifactorial nature of seafood composition. Many variables are thus combined that may influence the bioavailability of nutrients like omega-3 fatty acids or antioxidants: co-presence of other nutrients, metabolic status of the individual, and conditions of seafood preparation. Understanding these interactions requires complex experimental designs and sophisticated data analytical techniques that may not always be feasible or available.

Although the integration of omics technologies provides a depth of insight, some challenges are associated with data management and interpretation. Large datasets resulting from genomics, proteomics, and metabolomics studies need the involvement of advanced bioinformatics tools for analysis, which are very difficult to integrate with conventional biochemical data. In addition, the cost and skill required for such technologies are high, so it becomes inaccessible to most people, especially in resource-constrained settings.

1.8. FUTURE PROSPECTS OF SEAFOOD BIOCHEMISTRY

New technologies, together with a rising understanding of complex biochemical processes controlling seafood quality and nutritional value, place the seafood biochemistry field on a forward trajectory. Its future offers some possible breakthroughs and innovations that can change the way seafood is perceived and utilized and also how global challenges related to nutrition, food security, and sustainability are addressed.

1.8.1. Potential Breakthroughs and Innovation

New bioactive compounds and functional ingredients will be the most promising areas of future breakthroughs in seafood biochemistry. This is due to the constant research into health benefits from several compounds, such as omega-3 fatty acids, peptides, and antioxidants found in seafood; therefore, many expectations exist for the formation of new functional foods and nutraceuticals. Such products may be designed to fight cardiovascular diseases, memory decline, or other inflammatory conditions. It involves identifying and synthesizing certain compounds that can be put into supplements or used to fortify specific types of food that could deliver a more focused health benefit, thereby further extending the role of seafood in public health beyond just nutrition (Figure 1.3).

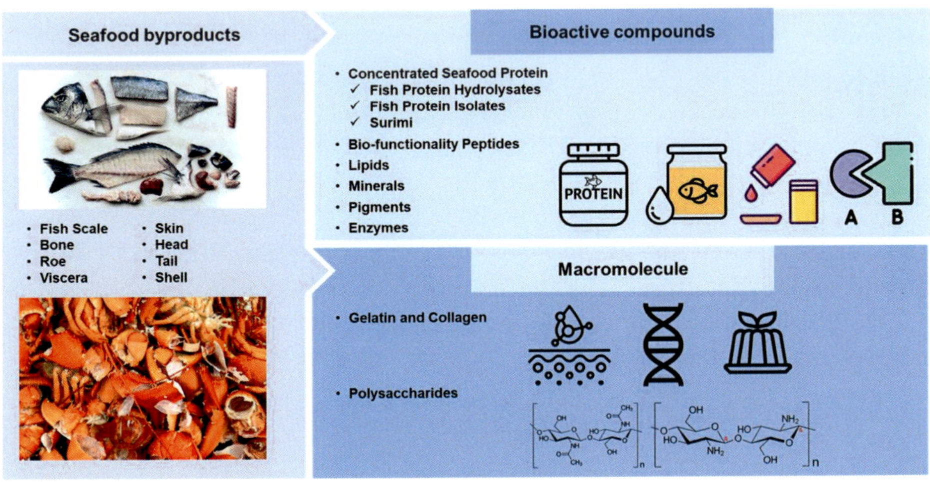

Figure 1.3. *A graphical representation of the seafood/fish/nutrient/bioactive compound.*

Source: Roy et al., (2023).

The omics technologies are probably going to affect seafood biochemistry to a great extent in the future. More detailed knowledge of the genetic, protein, and metabolic profiles of marine organisms is possible through these technologies. For instance, genomics can contribute to the identification of those genes that are connected with desirable characteristics such as high levels of omega-3 fatty acids or resistance against environmental stress. These findings can then be applied in aquaculture to breed species that are not only more nutritious but also hardier against varying environmental conditions. Similarly, both proteomics and metabolomics will yield information about protein and metabolic changes occurring during seafood processing and storage that can then be used to develop new preservation techniques and increase product quality.

Another forthcoming innovation is the use of biotechnology and synthetic biology in the production of seafood. These technologies have the potential to change the nature of industrial production concerning growing lab-grown seafood or enhancing traditional seafood species. Genetic engineering can also be used to enhance the nutritional value of seafood. For example, increasing levels of beneficial fatty acids or reducing harmful substances, like mercury. Lab-grown seafood, a product of cellular agriculture, is a sustainable and ethical alternative to wild-caught or farmed seafood, potentially taking pressure off overfished populations and limiting environmental impacts.

1.8.2. Areas for Further Research

While innovation in seafood biochemistry is an exciting potential, there are several gaps to be addressed. One important area pertains to the impact of environmental change, particularly climate change, ocean acidification, and pollution on the biochemical composition of seafood. With constant changes going on in the world's environment, there is a paramount need to concentrate on the quality, safety, and nutritional value of these seafood products in shifting environmental conditions. It could therefore contribute to the development of mitigating strategies for the negative impacts of environmental stressors on supplies of seafood, hence ensuring that they remain long-term and resilient.

Seafood allergens and food safety are equally crucial areas. The consumption of seafood has continued to increase globally; thereby, the rate of seafood allergies and concerns of foodborne illness also increased. Better studies of the prevalence of allergens in different seafood species and better diagnostic tools and treatments for seafood allergies are therefore required. The sources of various contaminants of seafood i.e., heavy metals, microplastics, and persistent organic pollutants leading to bioaccumulation, need to be deciphered for the development of proper monitoring and mitigation measures. Realizing the potential of underutilized and novel marine resources also requires more research. These marine organisms are largely unexplored for their biochemical composition and potential health benefits. This can provide new sources of nutrition, bioactive compounds, and even pharmaceutical ingredients. On the other hand, it is possible that investigating the biochemistry of these organisms will also lead to the identification of new compounds that have some specific features and open up absolutely new avenues for food and pharmaceutical applications.

Finally, there is a growing need in the area of consumer-oriented studies in seafood biochemistry. This would involve perception studies on consumers' knowledge of seafood quality, safety, and nutritional value, together with drivers of purchasing decisions. It would also help in developing improved communication and education programs to better promote the health benefits of seafood consumption while allaying concerns about its safety and sustainability.

1.9. ETHICAL CONSIDERATIONS IN SEAFOOD BIOCHEMISTRY RESEARCH

One of the major ethical dilemmas regarding research in seafood biochemistry is the source of the seafood. Overfishing and depletion of marine resources are among the pressing global issues, with ecological, economic, and social consequences. A researcher in seafood biochemistry or seafood-related firms must take into consideration the sustainability of the seafood used for studies or contents in the final products. This will mean choosing sources that do not contribute to the depletion of endangered species or alter marine ecosystems.

Another ethical concern is the impact of seafood harvesting on local communities. Indeed, in most parts of the world, local communities are greatly dependent upon fishing for their livelihood and food security. Unsustainable practices driven by high demand and weak regulation result in overfishing and deplete fish stocks, thereby making those communities vulnerable. Organizations and researchers need to move through these challenges by ensuring that the supplies are based on sustainable fishing methods and that sourcing will not have an impact on the livelihoods of local fishers or seafood availability in local populations.

There is also the issue of bycatch, where non-targeted species are caught unconsciously in the process of catching fish. Bycatch includes endangered species, juvenile fish, and other kinds of marine life. All these contribute to ecological imbalance and loss of biodiversity. In this regard, ethical sourcing would have to mean that methods and technologies that reduce bycatch have to be implemented and that seafood farming and trade will not adversely impact non-target species or contribute towards a loss in biodiversity.

1.9.1. Ensuring Transparency and Integrity in Research

Aside from sourcing, other ethical questions in seafood biochemical research pertain to the integrity and transparency of the research. Research demands openness. In this regard, this helps to engender trust from a public that is most definitely going to act on the research findings, which should be credible and reproducible. A researcher is under obligation to declare any kind of conflict of interest, especially when studies are funded by organizations that have commercial interests in the outcome. This transparency has the advantage of preventing biases that might influence the research findings by allowing the scientific community and the general public to look critically at the validity and reliability of the results.

It also involves carrying out all experiments and research studies with the highest possible degree of scientific rigor and integrity. This includes the proper design of experiments, the use of appropriate controls, and the application of accurate analytical methods that have been validated. Finally, ensuring good ethical use of animals in research, including making sure that the use of marine organisms is justified, humane, and in compliance with procedures for ethical use.

Data management and reporting are critical parts of ethical research. The scientist has to ensure that data is recorded accurately, securely stored, and made available for verification or replication by fellow scientists. Selective reporting of results, manipulation, exaggeration for sensationalism, or commercialism lowered the integrity of research with possibly harmful consequences, especially wrong information concerning safety or nutritional seafood benefits.

Chapter 2

Composition of Seafood

Contents

INTRODUCTION

Seafood is among the most diverse and nutrient-rich foods in the world. It entails a huge number of species: fish, crustaceans, mollusks, and other marine organisms. Seafood belongs to the group of very complex mixtures of biochemical components responsible for its nutritional value, flavor, and quality generally. Understanding seafood composition is very important not only to consumers but also to industries related to food processing, nutrition, and public health.

Nutritional sea food components include macromolecules like proteins, lipids, and carbohydrates and the essential micronutrients like vitamins and minerals. All of these play key roles in maintaining health and preventing diseases. For example, seafood delivers high-quality proteins for muscle growth and repair, and omega-3 fatty acids in most species decrease the risk of heart diseases and improve the brain function. This food is also an excellent source of some significant vitamins and minerals, specifically vitamin D, B12, iodine, and selenium, which are involved in different physiological processes.

The sensory properties of seafood i.e., flavor, texture, and aroma, are also a function of composition. These properties are determined by the particular types of proteins, lipids, and other compounds present and their respective proportions, which may vary enormously from one species to another and even within the same species because of variables such as habitat, diet, and processing methods. No less important is the need to determine this variability for many applications related to consumer preference, market value, and the control of food safety and quality. Moreover, it contains bioactive compounds like antioxidants and peptides that add a new dimension to seafood nutrition for possible health benefits beyond basic nutrition. In addition, seafood can accumulate several contaminants like heavy metals and persistent organic pollutants, some of which pose health risks and point out careful monitoring and regulation.

2.1. NUTRITIONAL COMPONENTS OF SEAFOOD

Seafood contributes to one of the richest sources of key and essential nutrients. It offers an overall good mix of macronutrients that are critical in keeping healthy and performing diverse bodily activities. Major macronutrients basically found in seafood include proteins, lipids, and carbohydrates; each has its peculiar contribution to the nutritive value of these foods.

Proteins are one of the most important macronutrients found in seafood and constitute a major part of the dry weight. In general, the protein content varies widely across species and may range from 15% to 25%. These proteins are of high biological value, meaning they contain all the essential amino acids required by the human body. These amino acids are largely responsible for muscle repair and growth, immune functions, and the production of hormones and enzymes. In addition, seafood proteins are well-tolerated and easily digestible, making them an excellent dietary choice for all age categories.

Another important component is the lipids present in seafood, particularly long-chain omega-3 polyunsaturated fatty acids. Prominent among them are eicosapentaenoic acid (EPA) and docosahexaenoic acid (DHA), mainly found in fatty fish such as salmon, mackerel, and sardines. These fatty acids are considered essential for brain health, cardiovascular function, and the regulation of inflammatory responses. Compared to most other sources of dietary fat, omega-3 fatty acids from seafood have been highly investigated and linked with numerous health benefits, including a reduced risk of heart disease, improved cognitive function, and possible anti-inflammatory effects.

Compared to proteins and lipids, seafood contains relatively small amounts of carbohydrates. Generally, carbohydrate content is less than 5%, primarily composed of glycogen in muscles and some free simple sugars. Although carbohydrates are only minor components of seafood, they still contribute to its energy value, especially during storage and processing. For example, glycogen can enhance the sweetness and texture of certain mollusk species (Figure 2.1).

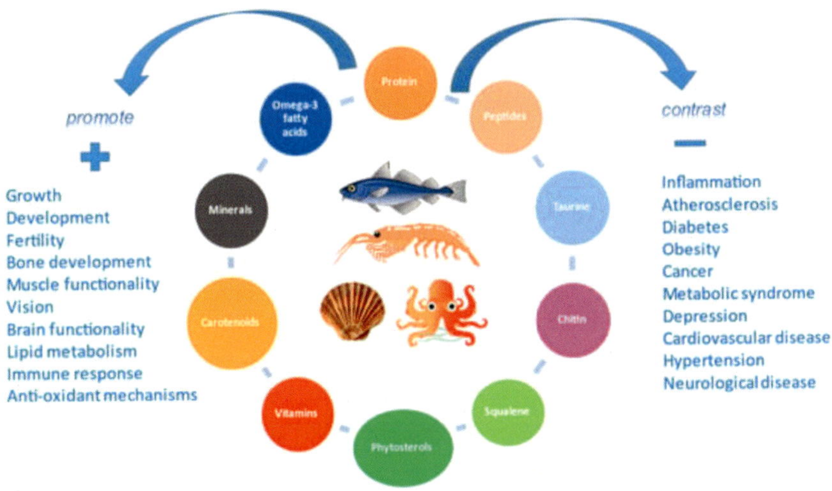

Figure 2.1. *Bioactive compounds present in various parts of fish.*

Source: Durazzo A, Di Lena G, Gabrielli P, Santini A, Lombardi-Boccia G, Lucarini M. Nutrients, and Bioactive Compounds in Seafood: Quantitative Literature Research Analysis. Fishes. *2022; 7(3):132. https://doi.org/10.3390/fishes7030132.*

2.1.1. Importance of Seafood in Diet

The importance of seafood in the diet cannot be overemphasized, considering their contribution to high-quality proteins and beneficial fats. Proteins found in seafood are not only complete but also lean, having very low saturated fat, which is a major consideration for heart health. Since it is a lean protein source, seafood is considered ideal for those wanting to keep slim or maintain a heart-healthy diet.

Seafoods are rich in omega-3 fatty acids which the human body is unable to produce in sufficient amounts and which have, therefore, to be consumed through food. The American Heart Association and other health organizations recommend at least two servings per week of fatty fish (Kris-Etherton, Harris, and Lawrence J. Appel, 2003). This recommendation is based on evidence that regular consumption of seafood rich in omega-3s lowers blood pressure, reduces triglycerides, and decreases the risk of heart arrhythmias, as well as slows the growth of arterial plaque.

In addition to proteins and lipids, seafood is a good source of many micronutrients of health importance. It is an excellent source of vitamins, in particular the B vitamins, including vitamin B12, which plays a crucial role in nerve function and in the formation of DNA and red blood cells. Seafood is also among the few natural food sources that provide vitamin D, essential for healthy bones and immune function. Besides, sea foods are good sources of essential minerals, wherein most species contribute substantial quantities of iodine, selenium, zinc, and iron. These major trace elements act on metabolic enzymes as cofactors, on antioxidant defenders, or in thyroid hormone metabolism.

The health benefit of seafood reaches beyond reducing the risk of chronic diseases. They are associated with improved cognitive development in infants and children; hence, it is recommended as part of the diet for pregnant and breastfeeding mothers. The DHA content in seafood is essential for brain development, resulting in better visual and cognitive outcomes.

It is, however, important to reiterate that whereas there are clear health benefits from the intake of seafood, there are considerations in relation to contaminants, particularly mercury, associated with some species of fish.

2.2. LIPIDS IN SEAFOOD

The main types of lipids in seafood include triglycerides, phospholipids, and sterols, which hold special functions in the sea organisms and human beings' health (Cardoso, Afonso, and Bandarra, 2016). Omega-3 PUFAs are particularly well-known for their significant contribution to health; examples are EPA and DHA.

2.2.1. Types of Lipids Present

Triglycerides are the main lipid class in fish, associated with the main form of energy storage. Triglycerides vary by fatty acid composition, which predominantly affects their nutritional value. Most marine species are rich sources of omega-3 fatty acids, mainly EPA and DHA. Fatty fish like salmon, mackerel, herring, and sardines are richly endowed with these omega-3 fatty acids. These omega-3 fatty acids are essential, meaning they cannot be adequately synthesized by the human system and therefore have to be ingested in the diet.

The other important type of lipids from seafood are phospholipids. The cell membranes, integral to these lipids, play a vital role in the maintenance of cellular structure and function. In seafood, phospholipids also contain omega-3 fatty acids, particularly DHA. DHA is very vital for the health of the brain and the eye. Phospholipid forms enhance their bioavailability, making these fatty acids more accessible to be used by the body.

Sterols, like cholesterol, are also found in fish and seafood. While cholesterol often connotes negative ideas about cardiovascular health, it is, in fact, an important molecule to the body, involved in the construction of cell membranes and in the synthesis of hormones. Seafood, in general, tends to have relatively moderate levels of cholesterol compared with other animal products, and the benefits of omega-3 fatty acids found in seafood generally outweigh any potential concerns associated with cholesterol intake (Figure 2.2).

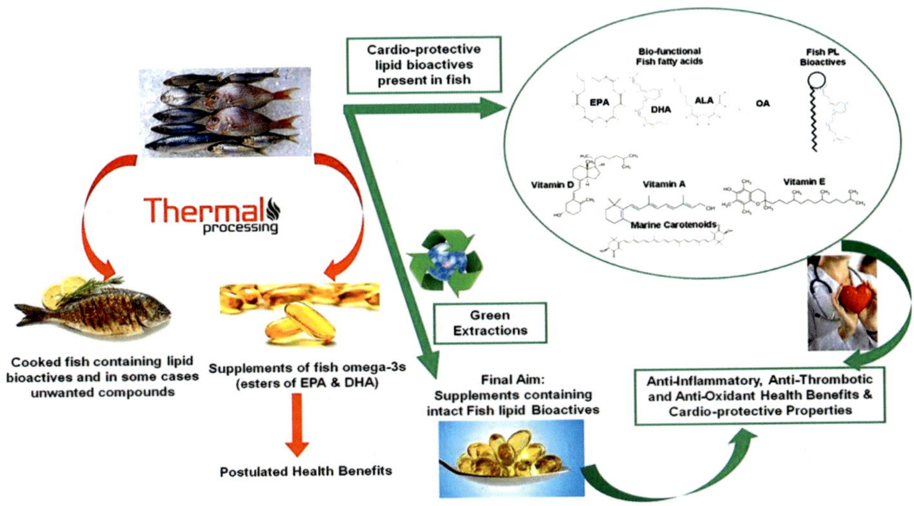

Figure 2.2. *Fish lipid bioactives.*

Source: Tsoupras et al., (2022).

2.2.2. Impact of Seafood Lipids on Human Health and Nutrition

The role and effect of seafood lipids on health and nutrition are very deep and wide-ranging. Omega-3 fatty acids play an important function in cardiovascular health, decreasing triglyceride levels, blood pressure, the risk for arrhythmias, and preventing the formation of arterial plaques. The anti-inflammatory activities of the omega-3 fatty acids are also useful in reducing the risk of chronic diseases like rheumatoid arthritis and inflammatory bowel disease.

Besides cardiovascular benefits, DHA is also needed for brain and eye development and functioning. Enough intake of DHA is particularly crucial during pregnancy and early childhood when it supports neural development and cognitive function. Studies have shown that DHA-rich seafood consumption by the mother may bring about better outcomes regarding visual and cognitive abilities in infants. At the other end of life, DHA is associated with reduced risk for neurodegenerative diseases, including Alzheimer's, an important constituent in the diet in a bid to sustain cognitive health. Seafood phospholipids also help in reaping these benefits by increasing the bioavailability and absorption of omega-3 fatty acids. The omega-3s of seafood origin are more effectively utilized by the body compared to sources like plant-based alpha-linolenic acid, which involves conversion to EPA and DHA. As expected, cholesterol and other sterols are consumed in seafood; however, associated intake is typically modest. Omega-3 fatty acids' benefits to lipid profiles likely offset potential negative effects associated with dietary cholesterol intake. Some risks associated with contaminants like mercury are predominantly found in predatory fish species, such as sharks and swordfish. Mercury binds with proteins of fish muscle tissues, and higher lipid contents at times correlate with increased levels of mercury. However, in most cases, the benefits derived from eating omega-3-rich fish outweigh the risks, especially when consumers choose lower-mercury options like salmon and sardines.

2.3. PROTEINS IN SEAFOOD

Seafood proteins exhibit superior quality in that they have a complete profile of amino acids and high digestibility, making it an excellent source of essential nutrients to maintain and build body tissues and support metabolic functions for energy.

2.3.1. Types of Proteins and Their Functions

The types of proteins in seafood can be broadly classified into three major groups: structural proteins, sarcoplasmic proteins, and connective tissue proteins, performing diverse functions in the organisms.

- **Structural Proteins (Myofibrillar Proteins):** These are very abundant proteins in seafood. They play a crucial role in muscle contraction and structure. It is composed of myosin and actin, which form the contractile apparatus of muscle fibers. Myofibrillar proteins contribute to the texture and firmness of seafood flesh; therefore, they are of primary importance in the sensory quality and nutritional value of seafood products. They also include some of the most important amino acids involved in muscle protein synthesis and repair, like leucine, isoleucine, and valine.
- **Sarcoplasmic Proteins:** This class of protein includes enzymes and other proteins of the muscle cytoplasm. It will contain enzymes such as creatine kinase, lactate dehydrogenase, and many other metabolic enzymes. It represents metabolic processes like energy production and glycolysis. Sarcoplasmic proteins also contribute to the flavor and color of seafood; it contains pigments like myoglobin, which may have a bearing on the flesh color of fish.
- **Connective Tissue Proteins:** Collagen is the major connective tissue protein in seafood, particularly found in species with more significant textures or in parts like skin and fins. It imparts both strength and elasticity to tissues. Although it comprises fewer essential amino acids than myofibrillar proteins, collagen and other connective tissue proteins are crucial for the physical properties seafood possesses, thus influencing its process characteristics and culinary uses.

2.3.2. Digestibility and Bioavailability of Proteins

Seafoods form a highly digestible and bioavailable source of proteins in comparison to most other sources. Several factors contribute to their high digestibility and bioavailability, most especially their amino acid makeup, the physical nature of the proteins, and the presence of some bioactive peptides.

Seafood proteins contain all nine essential amino acids in a pattern very near to human requirements. Such a high quality of the amino acid profile makes seafood an excellent source of protein with satisfying protein requirements, especially in countries where animal protein sources apart from fish are relatively inaccessible. Moreover, the presence of methionine, lysine, and tryptophan adds value to the nutritional properties of seafood by supporting a wide variety of physiological functions, from tissue repair to neurotransmitter synthesis.

Seafood generally has shorter and less dense muscle fibers compared to that of terrestrial animal origin, which makes them highly digestible. This basically makes it easier for the body to break down these proteins into their constituent amino acids and facilitates easier absorption and utilization. Compared with red meat, most seafood has relatively low levels of collagen, which helps in its tenderness and digestibility since collagen can be more resistant to breakdown.

Some bioactive peptides that impart health benefits can be released from seafood proteins during digestion. They may have antihypertensive, antioxidant, antimicrobial, and anti-inflammatory properties. For example, peptides from fish proteins have been found to be ACE-inhibitors, which maintain the blood pressure. Potential health benefits arising from these peptides add another layer to the nutritional value of seafood proteins beyond just their basic amino acid content.

2.3.3. Factors that Impact Digestibility and Bioavailability of Proteins

While seafood proteins generally exhibit high digestibility, some factors may have an impact on their bioavailability. Cooking methods, for example, may alter protein structure and hence impair digestibility. Overcooking may cause the denaturation of proteins, thereby reducing their digestibility and making some important amino acids unavailable. Conversely, some cooking modes, like steaming, can improve the digestibility by rupturing the muscle fibers (Bhat et al., 2021). This may further be affected by the presence of other dietary components, including fat and fiber, on digestion and absorption of proteins.

2.4. CARBOHYDRATES IN SEAFOOD

Although present, carbohydrates are not major constituents of seafood in comparison to proteins and lipids. The carbohydrate content in seafood is low and usually less than 5% when considering the general composition of seafood. Though at very minimal levels, carbohydrates in seafood play significant roles, more so in energy metabolism and on structural formation.

2.4.1. Presence and Types of Carbohydrates

Types of carbohydrates that may be found in seafood include simple sugars, polysaccharides, and glycogen. They are found in small portions of simple sugars, which include glucose and ribose, that act as simple basic sources of energy. In essence, the inclusion of these simple sugars forms part of many of the various cellular processes, for instance, the synthesis of nucleotides and nucleic acids which are requirements in cellular replication and repair. Some of the polysaccharides found in seafood are chitin, a major component of the exoskeleton of crustaceans and in some cell walls of a few marine organisms, including algae. Chitin is a long-chain polymer of N-acetylglucosamine, and thus it provides, as an important structural component, rigidity, strength, and protection to the exoskeletons. It is an important carbohydrate for the organisms themselves, as it is indigestible in its native form. Chitin, however, is processable into a variety of uses in food technology, pharmaceuticals, and water purification, as its derivative chitosan is biocompatible and non-toxic.

Glycogen, a polysaccharide that represents the principal stored form of energy in animals, is also found in seafood and is mainly of mollusk and shellfish origin. The glycogen content may vary by several orders of magnitude depending on species and time of harvest. For instance, mollusks such as oysters and scallops have a high content of glycogen, possibly the reason for their sweetness and texture. Glycogen is readily broken down into glucose units, thus offering an immediate source of energy in such animals when food is in low supply.

2.4.2. Role in Metabolism

Although seafood is not a substantial source of carbohydrates in human nutrition, the present carbohydrates do play roles in the metabolism of marine organisms and to some lesser extent in human nutrition. In the organisms per se, carbohydrates play a crucial role in various metabolic pathways. It acts as a reserve of energy that is readily mobilized in times of high energy demand, such as during muscle contraction or food deprivation. In terrestrial animals, glycogen also acts in the liver and muscles as a quick-release source of energy.

Even though the intake of carbohydrates from seafood is very low in humans, the carbohydrates present can still have an effect; notably, in the context of dietary fiber, chitin and its derivative, chitosan, act like dietary fiber. They may assist in the management of cholesterol and act as a prebiotic, providing gut health through the stimulation of beneficial gut bacteria. In fact, chitosan alone has been tested for its possible mechanisms of lowering cholesterol levels and causing weight loss by binding with dietary fats and preventing intestinal absorption.

This glycogen, therefore, renders some of the seafood species to contribute to their sensory properties, including taste and texture. The sweetness in these sea creatures, as established by their content of glycogen, makes them more favored than others in a culinary application. It can enhance the general palatability and enjoyment of fish dishes, hence increasing their liking among consumers.

While carbohydrates in seafood are not a key issue of dietary recommendations, their role in the metabolism and structure of marine organisms, as well as in some of the sensory qualities of seafood, cannot be ignored. For example, the rather low content of carbohydrates in seafood, together with its high content of proteins and omega-3 fatty acids, makes it quite an ideal constituent for low-carbohydrate diets that may have positive effects on weight management and metabolic health.

2.5. VITAMINS AND MINERALS IN SEAFOOD

Seafood is a good source of many vitamins and essential minerals, and these nutrients have some important bodily functions to keep people healthy and help prevent certain conditions and diseases. A variety of physiological activities involved in immune function, bone health, energy metabolism, and even neurological function are entrusted to key micronutrients. The most useful vitamins and minerals found in fish are vitamin D, vitamin B12, iodine, selenium, and omega-3 fatty acids. Each of these has its unique health benefits.

2.5.1. Key Vitamins and Their Health Benefits

The most prominent vitamin consumed from seafood, especially in fatty fish like salmon, mackerel, and sardines, is vitamin D. This is an essential nutrient for the assimilation of calcium in the body and is vital for maintaining bone health; it thus prevents disorders such as osteoporosis and rickets. Vitamin D is also involved in immune function, reducing the risk of infections, and may decrease the incidence of autoimmune diseases. Coming from the fact that sunlight exposure, which is a natural source of vitamin D, is particularly limited in most areas during winter months, seafood becomes an essential dietary source.

Specifically, shellfish and clams, along with fish, especially trout and salmon, represent an extremely rich source of vitamin B12. Vitamin B12 is crucial for the production of red blood cells and DNA synthesis, but it also plays an equally important role in neurological function. Deficiency in this vitamin will lead to anemia, fatigue, neurological disorders, and, worst of all, loss of memory or even dementia. Since the main source of this vitamin comes from animal products, seafood assumes a very important place for those who consume a low intake of other foodstuffs from animals.

2.5.2. Key Minerals and Their Health Benefits

Among the minerals, iodine, and selenium are particularly notable in seafood. Iodine is essential for thyroid function, as it is a critical component of thyroid hormones, which regulate metabolism, growth, and development. Seafood, especially seaweed and fish, is one of the best natural sources of iodine. Adequate iodine intake prevents thyroid-related disorders, such as goiter and hypothyroidism. Iodine deficiency is a common issue in areas where seafood consumption is low, leading to significant health problems, especially among pregnant women and children.

Selenium is another vital mineral found in seafood, particularly in Brazil nuts and seafood like tuna, sardines, and shrimp. Selenium has antioxidant properties, protecting cells from damage caused by free radicals. It also supports immune function and thyroid health. Adequate selenium intake is associated with a reduced risk of certain cancers and cardiovascular diseases. It plays a critical role in DNA synthesis and reproductive health as well.

In addition to these, seafood is also a good source of other essential minerals like zinc, calcium, and iron. Zinc is crucial for immune function, wound healing, and DNA synthesis. Calcium, found in smaller amounts in some seafood like sardines and salmon (with bones), is essential for bone health and muscle function. Iron, necessary for the formation of hemoglobin and oxygen transport in the blood, is present in seafood like clams, oysters, and fish, contributing to the prevention of iron-deficiency anemia.

2.6. ANTIOXIDANTS IN SEAFOOD

Antioxidants are compounds that inhibit oxidation, a chemical reaction that may produce free radicals. Those free radicals could further lead to cell damage. The antioxidants of seafood improve its nutritional value and exert important health benefits. They aid in the protection of the body against oxidative stress, which is related to a large number of chronic diseases, from heart diseases and cancer to neurodegenerative disorders. The major sources of seafood antioxidants include carotenoids, tocopherols, and certain polyphenols and also some peptides and enzymes.

2.6.1. Types of Antioxidants and Characteristics

Probably the best-known seafood antioxidant group is carotenoids, especially astaxanthin and beta-carotene. Astaxanthin is a red-orange pigment found in the following crustaceans: shrimp, crab, lobster, salmon and trout. This carotenoid is noted for its extremely powerful antioxidant activity, believed to be far greater than other well-known antioxidants, such as vitamins E and C. Astaxanthin has an antioxidant capacity that protects from oxidative damage to cells by neutralizing free radicals, which may lead to decreased inflammation and actually enhanced immune function.

Beta-carotene is another carotenoid antioxidant that serves as a precursor to vitamin A. Major sources of beta-carotene include seafood, particularly shellfish and some fish. Other than having antioxidant activity, it plays a major role in vision, immune function, and skin health. This conversion of beta-carotene into vitamin A is vital since vitamin A is important in maintaining healthy vision, immune response, and cellular communication.

Other seafood, in particular fish oils and a few fatty fish, also contain another group of chemical compounds known as tocopherols, of which is vitamin E. Vitamin E is a fat-soluble antioxidant and interferes in cell membrane oxidative damage due to lipid peroxyl radical scavenging. Such protection is essential in maintaining the integrity of cell membranes and preventing lipid peroxidation, which can cause cell damage and, hence, inflammation.

While polyphenols are more generally associated with plant-based foods, they are also found in some types of seafood, particularly those that eat algae or plankton rich in these compounds. Levels of polyphenols can therefore be accumulated in fish such as mackerel and sardines that are higher up in the food chain and whose diet consists mainly of smaller marine organisms. These compounds thus participate in the antioxidant activity of seafood and help counteract oxidative stress and inflammation.

Besides these well-known antioxidants, seafood also contains some unique bioactive peptides and enzymes that are capable of antioxidant activity. For instance, some peptides isolated from fish proteins have been proven to show strong antioxidant activity in inhibiting the process of lipid oxidation and in stabilizing free radicals. Equally important, other enzymes found within marine organisms, such as superoxide dismutase and glutathione peroxidase, are involved in critical ways in body defense against oxidative stress by catalyzing the catabolism of injurious superoxide radicals and peroxides.

2.6.2. Benefits of Antioxidant Consumption

Seafood is rich in antioxidants that provide a variety of health benefits associated with the reduction of oxidative stress and inflammation, two major contributors to chronic diseases. Antioxidants neutralize free radicals, thereby protecting cellular elements from oxidative damage, especially DNA, lipids, and proteins. Most importantly, in avoiding those diseases where the main mechanism at play is oxidative stress, such as cardiovascular disease, cancer, and neurodegenerative disorders like Alzheimer's and Parkinson's disease, this becomes especially crucial.

Among the most important causes of oxidative stress is atherosclerosis, or hardening and narrowing of the arteries by deposited oxidized LDL cholesterol. Vitamins E and astaxanthin are some of the antioxidants that could inhibit the oxidation of LDL cholesterol, hence decreasing the possibility of plaque formation and cardiovascular accidents in general, such as myocardial infarction and strokes.

Concerning cancer prevention, antioxidants work to protect cells from potential gene mutation resulting from oxidative damage. Stabilizing free radicals, the antioxidants of seafood may help to prevent possible mutations in DNA that cause cancerous growths. Astaxanthin and some other antioxidants under study have recently gained significant interest for their potential role in enhancing immune response and in the inhibition of the growth of cancer cells, thus being of great value as a component of a diet that would prevent the condition.

Another critical difference is related to the neuroprotective benefits of seafood antioxidants. Oxidative stress plays a central role in the process of senescence and development of neurodegenerative diseases. On this respect, DHA, an omega-3 fatty acid with antioxidant properties, may bind with carotenoids in protecting neuronal cells against oxidative damage and sustain cognitive function throughout life, possibly slowing down age-related cognitive decline and neurodegenerative diseases.

2.7. WATER CONTENT IN SEAFOOD

Water is a major constituent of seafood, having a huge effect on its texture, quality, and culinary performance. This variability of water content in seafood may be enormous among different species and even within the same species due to variables like age, size, and habitat. As a rule, seafood is characterized by its high water content, which usually ranges from approximately 60% to over 80% of the total weight. The latter is typical for the best sources of meat and the result of the fact that sea creatures are accustomed to live in an aqueous medium and often develop additional mechanisms that regulate water balance. Nevertheless, numbers can differ greatly depending on the species and seafood type.

For instance, lean fish like cod, haddock, and tilapia usually have over 80% water content. These species have less fat and hence rely more on water for buoyancy and tissue structure. On the other hand, fatty fish such as salmon, mackerel, and tuna contain less water, normally in the range of 60%–70%, since they have a higher percentage of fat content that replaces part of the water in their tissues. It is not only about the nutritional profile of those fish being different due to this difference in water content; even the texture and culinary uses are affected.

Water content levels in shellfish, such as shrimp, crabs, and lobsters, may vary, and it can be due to their habitat and physiology. For example, crustaceans from more saline environments may have different in-water retention capabilities than those that are in freshwater. Mollusks, for example, clams and oysters, have a high content of water since they are soft-tissue-structure, colorless inclusions filled with cells of water. In addition to high water content, these species will also have a characteristic texture and flavor.

2.7.1. Implications for Texture and Seafood Quality

The water content in seafood has serious implications for texture and quality. In most cases, high water content translates into tenderness and juiciness, an attribute viewed with much importance in many culinary applications. For example, the flakiness of a well-cooked fish fillet is partly due to the high water content in the muscle fibers, which creates a delicate and succulent texture when properly cooked. This feature is particularly valued in lean fish species, where the water content somewhat offsets the low levels of fat, making it more palatable and pleasing to eat.

On the other hand, water content can also create problems during processing and storage. Seafood that is highly water-intensive, has a greater tendency to become spoiled or to experience the growth of microorganisms, as water is the perfect medium for bacteria and other micro-organisms. This calls for proper handling, refrigeration, and preservation methods in the process of maintaining the quality and safety of seafood products. Freezing is generally employed to preserve high-water-content seafood products, so it needs to be done with the utmost care, as problems like ice crystal formation may cause the destruction of cellular structure and hence loss of texture when they are thawed.

This variability in water content also affects the weight and yield of seafood products, which is of great importance commercially and culinarily. Variations in water content might lead to more loss in weight during cooking because of water loss, hence affecting the size of portions and the texture at the end. This is of special consideration for creatures like shrimp and scallops that can shrink during cooking to a remarkable degree if not handled properly. Techniques such as brining or marinating could, therefore, be used to help in the retention of moisture and flavor.

Moreover, the water content determines even the apparent freshness and quality of seafood. Fresh fish has a firm, elastic texture, partly because it retains water within the muscle tissues. If seafood becomes old or is stored too long, the exudation may lead to a loss of firmness, and the product will appear dry and less fresh. This can usually be associated with other forms of quality deterioration, such as changes in color and odor.

In addition to texture and quality, seafood may differ nutritionally based on its water content. For example, species with lower water content and higher fat content will be much more concentrated about nutrient content, such as omega-3 fatty acids and fat-soluble vitamins. That said, the nutrient profile of seafood that has a higher water content is basically lower in density but remains one of the primary sources of good-quality protein and major minerals.

2.8. CONTAMINANTS IN SEAFOOD

Seafood contaminants can broadly be classified into chemical, biological, and physical (Figure 2.3):

- **Chemical Contaminants:** These are heavy metals, persistent organic pollutants, and toxins. Heavy metals stand for one of the most important groups of chemical contaminants in seafood. Some of the most dangerous heavy metals include mercury, lead, cadmium, and arsenic. Mercury is probably the best-known heavy metal contaminant of seafood, particularly in its form as methylmercury. It can bioaccumulate in the tissues of marine organisms, especially in large predatory fish like shark, swordfish, king mackerel, and some types of tuna. Persistent organic pollutants are industrial chemicals like PCBs and dioxins that persist in the environment for long periods. These contaminants get stored in the fatty tissues of fish and other marine life. Furthermore, seafood can sometimes forage for toxins from harmful algal blooms, such as saxitoxins, ciguatoxins, and domoic acid—all responsible for shellfish poisoning.
- **Biological Contaminants:** This group includes bacteria, viruses, and parasites. Certain pathogenic bacteria are known to be hosted by raw or undercooked seafood, particularly in shellfish. These come in the form of Vibrio species, Salmonella, and Listeria. These bacteria may cause a foodborne illness or even, in serious cases, grow into severe infections (Papadopoulou et al., 2007). Other very important concerns are viruses, chiefly norovirus and hepatitis A, particularly in shellfish that derive from contaminated waters. Other parasites have the anisakid nematodes and tapeworms that infect human beings when seafood is consumed raw or undercooked, resulting in gastrointestinal symptoms and sometimes even serious systemic health problems.
- **Physical Contaminants:** These are the least common, but include plastic wastes, microplastics, and other foreign bodies. A more recent concern is microplastics because, if ingested by sea organisms, they eventually enter the human food chain. While health effects from ingestion of microplastics are under investigation, there is a concern that they might either leach chemicals or cause physical damage to the digestive tract.

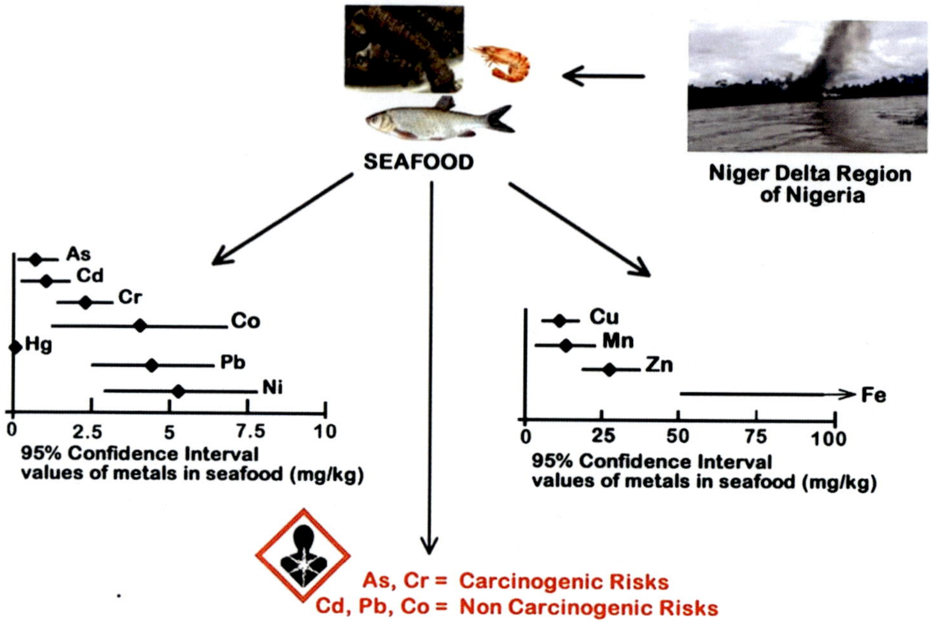

Figure 2.3. *Graphical illustration of heavy metal contents of seafood from the Niger Delta region of Nigeria.*

Source: Umeoguaju et al., 2023: Toxicology Reports, Volume 11: https://doi.org/10.1016/j.toxrep.2023.06.011.

2.8.1. Health Risks Associated with Contaminants

The health risks associated with seafood contaminants differ quite variably, depending on the contaminant type and dose. One of the heavy metals is mercury, which includes serious neurotoxic impacts, especially on women at pregnancy stages, fetuses, and children at young ages. Chronic exposure to mercury may result in cognitive development impairment and neurological deficit. Equally, lead exposure causes developmental, especially growth, delays and cognitive malfunctioning in children. Cadmium and arsenic inflict damage on the kidneys and lead to increased cases of cancer.

Persistent organic pollutants, such as PCBs and dioxins, are endocrine disruptors and have been associated with a variety of adverse health effects, including suppression of the immune system, reproductive and developmental problems, and increased incidence of certain cancers. Since the human system cannot easily rid itself of these fat-soluble compounds, there can be cumulative exposure with possible long-term health effects.

Biological contaminants in seafood can cause acute foodborne illnesses. These bacterial pathogens might then cause nausea, emesis, diarrhea, and in severe cases of infection, even septicemia or meningitis. Most importantly, viral infections, mainly caused by norovirus, are the most prevalent cause of the outbreaks of gastroenteritis linked with the consumption of seafood. Such infections are very dangerous to any immunocompromised individual, elderly person, or young child. Infection by parasites can result in a myriad of symptoms, ranging from light uneasiness to life-threatening conditions.

Different types of symptoms, ranging from light uneasiness to life-threatening conditions, are caused by infection due to parasites. Marine biotoxins such as those responsible for paralytic, neurotoxic, and amnesic shellfish poisoning can cause a variety of symptoms: gastrointestinal lesions to severe neurological effects, even death in extreme cases.

Although contaminants in seafood are a central concern, these risks should be weighed against the nutritional benefits of seafood consumption. Different governments around the world monitor and regulate the level of contaminants in seafood to make sure they remain within safe limits. Consumers can reduce the risks by choosing seafood that is lower in contamination, such as smaller fish or farmed species, and by following guidelines on seafood consumption as outlined by health authorities.

2.9. FACTORS AFFECTING COMPOSITION OF SEAFOOD

The constitution of seafood is an interplay between environmental conditions, genetic makeup, and species-specific characteristics. These different factors comprise the nutritional profile, flavor, texture, and quality in seafood and are important for commercial and scientific purposes within the production of this source of food.

Among the environmental variables, water temperature stands out as the most critical. The various temperatures offer different metabolic rates for the marine organisms in regards to growth, reproduction, and biochemical processes. For instance, fish from colder waters contain higher amounts of unsaturated fats, mainly omega-3 fatty acids, than those in warmer waters. This adaptation helps to keep the membrane of the cell's fluid at lower temperatures. Thus, fish like salmon, mackerel, and herring are very good sources of omega-3 fatty acids, making these fish particularly useful in human nutrition.

Salinity and water quality also influence seafood composition significantly. Marine organisms associated with different salinity conditions of water regulate their salt and water balance internally, which could have effects on their nutrient composition. For instance, seafood from saltier waters may differ in their mineral content from those obtained from less saline environments. In addition, water pollution and certain contaminants, including heavy metals and POPs, can be bioaccumulated in the organisms and hence affect the safety of the seafood for consumption. It is, therefore, most important to monitor water quality and environmental contaminants for the healthiness of seafood.

The nutrient availability within the aquatic environment, including the existence of phytoplankton and other food sources, has an impact on the nutrient content of seafood. When fish consume omega-3 rich phytoplankton or smaller fish already containing these beneficial fatty acids, it accumulates in the predator fish and makes them more nutritionally good. As such, the presence of some algae may cause the bioaccumulation of certain bioactive compounds like astaxanthin in shellfish and crustaceans, thereby influencing their content of antioxidant compounds.

2.9.1. Genetic Factors and Species Variability

Another major determinant of seafood composition is its genetics. Large variations exist among the nutritional profiles of different species and strains of the same species with differences in their genetics. Genetic variability can cause differences in the proportion of muscle to fat tissue in fish, therefore providing variability in protein and lipid contents (Parlapani, 2021). This is particularly the case with aquaculture, where selective breeding can be done for desirable traits such as faster growth rate, higher content of omega-3 fatty acids, or better resistance to diseases.

Variability among species is exemplified through the wide array of textures, flavors, and nutritional profiles represented within the seafood types. For instance, while oily fish like salmon and sardines are higher in fat, thus serving as dense sources of omega-3 fatty acids, lean fish like cod and haddock are higher in protein but low in fat. The other category consists of shellfish such as shrimp and scallops, which have quite different flavors and textures partly due to differences in the content of amino acids and glycogen. These differences are of interest not only to culinary art but are of real importance to both dietary planning and nutrition.

Nutrient bioavailability can also be influenced by genetic diversity within species. For example, differences in fish muscle fiber composition may result in varying effects on digestion and absorption of proteins by the human body. Furthermore, some species are genetically more predisposed to accumulating higher levels of particular contaminants, such as mercury and PCBs, but this should also be taken into account when considering the relative safety of a range of seafood products.

Moreover, genetic factors interact with environmental conditions to further affect seafood composition. This will occur through the phenomenon of phenotypic plasticity, where organisms change physiology or behavior due to changing environments. For instance, fish increase their fat stores in response to colder water temperatures, a trait both genetic and environmentally adaptive. Understanding the nature of these interactions is critical in aquaculture, where it has been controlled to attain the full nutritional value and safety of cultured seafood.

Chapter 3

Post-Harvest Biochemical Changes in Seafood

Contents

INTRODUCTION

Once seafood has been harvested, several biochemical changes occur, impacting seafood quality, safety, and shelf life. These changes can be due to a combination of intrinsic factors i.e., the seafood enzymes and microorganisms, coupled with extrinsic ones such as temperature, oxygen availability, handling practices, and so forth (Sikorski, Kołakowska, and Burt, 2020). These biochemical processes are very important to understand from the point of view of the seafood industry and consumers, since they define the final product's sensory attributes and its nutritional value, besides ensuring its safety.

Among the most important biochemical processes initiated shortly after harvesting is autolysis, meaning the self-digestion of seafood tissues due to the action of endogenous enzymes. This may lead to proteolysis, lipolysis, and degradation of other macromolecules with changes in texture and flavor. Similarly, the oxidative processes mainly lipid oxidation can cause rancidity. In this way, changes can result in a sensory deterioration that negatively affects the quality and also the nutritional value of seafood. The role of microbial spoilage—the proliferation of microorganisms with the production of compounds causing spoilage which further complicates this post-harvest scenario, leading to potential risks to quality and safety.

Enzymatic reactions, either from the seafood itself or from microbial sources, are critical to the rate and fullness of these biochemical changes. Such reactions may lead to protein denaturation, carbohydrate degradation, and formation of off-flavors and odors—events that modify consumer perception and acceptance of seafood products.

These changes are managed and mitigated by various factors, ranging from temperature control to proper storage conditions, use of preservatives, and packaging technologies. This would reduce the rate or come to a complete standstill of the biochemical changes that result in quality deterioration and spoilage.

3.1. AUTOLYSIS IN SEAFOOD

Autolysis or self-digestion is a natural, biochemical post-harvest process of seafood, where the tissues and cells undergo decomposition because of their own, self-releasing lysosomal enzymes. Autolysis is a very important mechanism of quality degradation in seafood, mostly affecting texture, flavor, and general acceptability. A proper understanding of the mechanisms and the factor or factors that control this rate is quite essential for managing the quality and shelf life of seafood products.

3.2. AUTOLYSIS IN SEAFOOD: MECHANISMS

3.2.1. Mechanisms of Autolysis

The self-digestion process of autolysis in seafood begins immediately after it has been harvested and the organism starts to die, which means normal cellular activities come to an end. The result is the release into the cytosol of hydrolases from lysosomes, a certain kind of organelle containing digestive enzymes. These enzymes will have an initiation effect on cellular components by hydrolyzing them.

Proteases are one of the key classes of enzymes involved in the autolysis process by which proteins are broken down into smaller peptides and amino acids. This proteolytic degradation process gives the muscular tissue a change in texture and affects the water-holding capacity. Typical of this process in such conditions is the action of cathepsins, calpains, and collagenases depending on the species. For example, cathepsins are very active at the low pH values that often build up in seafood tissue after its death and are known to increase the breakdown of muscle proteins.

Lipases are some of the important enzymes involved in lipid hydrolysis, particularly in oily fish. This group of enzymes hydrolyzes triglycerides into free fatty acids and glycerol, which in turn will be oxidized to produce rancidity and off-flavors. The activity of lipases is, therefore, of paramount concern in high-fat content species since it might rapidly degrade seafood's sensory qualities. Nucleases digest nucleic acids into nucleotides and breakdown products. Not directly related to sensory impact, the activity of nucleases in turn can contribute to the general degradation of the cellular structure.

Autolysis also involves the digestion of the cell membranes and other structural components, causing cellular fluids to flow out, thus leading to a loss of their water-holding capacity. This causes a softer, less desirable texture in the seafood, often judged as a loss of freshness.

3.2.2. Factors Influencing Autolysis Rate

Factors that influence the autolysis rate in seafood include temperature, species-dependent enzyme content, the pH value, and the handling conditions.

Of all the factors, temperature has the most profound effect on autolysis. Generally, as the temperature rises, so does the rate of activity of the enzymes; hence seafood stored at higher temperatures will undergo autolysis more quickly. This explains why rapid chilling and cool storage are very important in reducing the rate of autolysis and in extending shelf life. Freezing has a huge potential to stop autolysis because of inactivation of the enzymes; however, some enzymatic activity may resume when it is thawed. Species-specific enzyme content is also a strong factor that could significantly influence the rate of autolysis. There are wide variations in the levels of endogenous enzymes between fish species, which are partly responsible for the differences in the rate of development of autolysis. Species with a high-fat content, like mackerel and herring, have highly active lipases, and therefore the rate of lipid breakdown or spoilage is quickened. On the other hand, lean fish like cod might undergo less prominent changes in autolysis concerning lipid degradation but still rather intense protein breakdown.

pH levels of seafood tissue can also affect enzyme activity. The natural drop in pH following death, often caused by glycogen breakdown into lactic acid, can activate certain enzymes such as cathepsins with optimal activities at lower pH levels. This acidification could contribute to faster muscle protein breakdown, especially collagen, and thus loss of texture and firmness (Nikoo, Benjakul, and Gavlighi, 2022).

Post-harvest handling practices, such as gutting, washing, and cleaning, can also affect autolysis. Delayed gutting or inadequate cleaning of the fish will release more enzymes from the gastrointestinal tract, which will increase the rate of autolysis. Moreover, poor handling of seafood causes cell and tissue disruption, which enables easier access of enzymes to substrates and thus enhances degradation.

Finally, the quality at the time of catching is also one of the very important factors. Seafood stressed, hurt, or otherwise jeopardized before catching might have faster autolysis due to higher amounts of tension-related enzymes or even cellular integrity compromise.

3.3. OXIDATIVE PROCESSES IN SEAFOOD

Oxidative processes largely contribute to the deterioration of seafood quality post-harvest; among these, the contribution of lipid oxidation is very high. Lipid oxidation will result in rancidity, which adversely affects flavor, odor, color, and nutritional value. The mechanism of lipid oxidation and its preventive and controlling measures are very important to maintain the quality and extend the shelf life of seafood products (Figure 3.1).

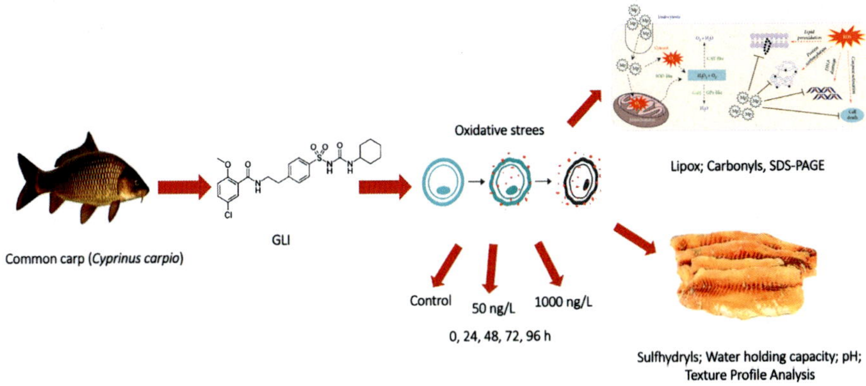

Figure 3.1. *Lipid and protein oxidation in seafood.*

Source: Martínez-Viveros et al., (2023).

3.3.1. Lipid Oxidation and Rancidity

Lipid oxidation in seafood concerns primarily the unsaturated fatty acids that are constituent parts of fish oils, more specifically omega-3 fatty acids like EPA and DHA. These are very easily oxidized because of their many double bonds, which offer a large number of sites vulnerable to oxidative attack.

The process of lipid oxidation is initiated by the formation of free radicals. Usually, this is initiated by exposure to light, heat, or metal ions. These formed free radicals react with the unsaturated fatty acids of the lipids to form peroxides of lipids. This step called the initiation phase, is followed by the propagation phase, whereby the lipid peroxides break down into secondary oxidation products such as aldehydes, ketones, and alcohols. It is these secondary oxidation products that provide the strong rancid off-flavors and off-odors of many of the oxidized seafood. Only when either all of the free radicals have been quenched or the substrate is exhausted is the chain finally terminated, but this may sometimes be too late to avoid some deterioration in quality.

Lipid oxidation has effects not only on the sensory qualities of seafood but also on its nutritional value. Oxidation can result in degradation of essential fatty acids and fat-soluble vitamins, particularly vitamins A and E, which are important in maintaining general health. In addition, some of the formed products from such an oxidation process, such as aldehydes, are known to be toxic; hence, their accumulation in food for human consumption may pose health risks.

3.3.2. Prevention and Control Measures

Lipid oxidation in seafood has to be prevented or controlled to guarantee its quality, safety, and nutritional value. From strategies applied with proper handling and storage to the application of antioxidants, to advanced packaging technologies, as well as temperature control, the means to this end are multifarious.

- **Temperature Management:** One of the most potent ways to prevent oxidation is through proper temperature management. Oxidation rates are temperature-dependent, so chilled or frozen seafood will experience the process at much slower rates. Freezing will be very effective because it not only inhibits enzyme activity but also reduces the availability of oxygen as a reactant for oxidative processes. However, freezing does not completely shut down oxidation; thus, it is very important to maintain a consistently low temperature.
- **Oxygen Control:** Oxidation can also be avoided by restricting the levels of oxygen that come into contact with lipids. The techniques most generally applied for these means in the seafood sector include vacuum and modified atmosphere packaging. Vacuum packaging removes air from the package, thereby reducing the level of oxygen, while modified atmosphere packaging replaces the air with a mixture of gases, generally nitrogen or carbon dioxide, which do not support oxidation (Raghavan and Kristinsson, 2014). It follows that such packaging can significantly extend the shelf life of seafood by reducing the rate of lipid oxidation.
- **Antioxidant Use:** Antioxidants added to seafood or in its packaging can block lipid oxidation processes. Some antioxidants, such as ascorbic acid (vitamin C), tocopherols (vitamin E), and synthetic compounds like butylated hydroxyanisole (BHA) and butylated hydroxytoluene (BHT), may inhibit the formation of free radicals or neutralize them once formed. Natural antioxidants resulting from herbs and spices have become popular due to the consumer preference for natural additives.
- **Light Protection:** More specifically, light, in particular ultraviolet light, can enhance the oxidation process of lipids through the generation of free radicals. Protecting seafood from light is thus another essential control measure in the processing chain. This can be achieved by packing it in opaque or UV-blocking material. Storage in the dark or coating with compounds that block light on packaging will further minimize photo-oxidation risks for seafood.

- **Avoiding Pro-Oxidants:** Pro-oxidants are metal ions such as iron and copper that can catalyze oxidative reactions. Their level within seafood and processing should be reduced as much as possible. This can be achieved using non-reactive packaging materials and equipment and adding chelating agents, which bind to metal ions and render them inactive.
- **Freshness and Hygiene:** Reduction of lipid oxidation can also be done by generally maintaining the freshness and hygiene of seafood. Fresh seafood will have undergone fewer initial oxidative processes, and good hygiene practices will introduce fewer catalyzing contaminants.

3.4. ENZYMATIC REACTIONS IN SEAFOOD

Enzymatic reactions form the core of postharvest biochemistry in seafood, influencing its texture, flavor, and overall quality. Certain enzymes that constitute these enzymatic reactions are naturally found in marine organisms and become active after harvest when the organisms die and their cellular systems stop functioning normally. The principal participant enzymes in this connection are proteases, lipases, and enzymes linked to the degradation of nucleotides and carbohydrates. Knowing the functions of these enzymes and their impacts is very important in managing seafood quality from catch to consumer.

3.4.1. Key Enzymes Involved

- **Proteases:** They are a group of enzymes that break down proteins into smaller peptides and amino acids. They are probably the most important enzymes in seafood, which have big impacts on both texture and flavor. On a broad level, proteases can be divided into endoproteases and exoproteases, based on their action on protein molecules. The endoproteases include cathepsins and calpains, which cleave peptide bonds within protein chains, while the exoproteases include aminopeptidases, which act from the ends of peptide chains. These enzymes show maximum activity in the early post-mortem changes and are responsible for the softening of seafood flesh, a process that is desirable in its early stages but undesirable if it goes too far.
- **Lipases:** It catalyze the hydrolysis of triglycerides into free fatty acids and glycerol. This reaction is important in lipid-rich fish species. As a result of the action of lipases, free fatty acids could be released and would have a higher tendency to become oxidized, developing rancidity and off-flavors. Thus, the activity of the lipases

becomes a critical factor in the sensory quality of seafood, so much so in omega-3-rich species like salmon, mackerel, and sardines.

- **Nucleotide-Degrading Enzymes:** Examples include ATPases and nucleotidases, which hydrolyze adenosine triphosphate and other nucleotides post-mortem to produce IMP or inosine monophosphate. IMP is considered generally responsible for the umami flavor of fresh seafood, although further breakdown to hypoxanthine might cause bitterness and unpleasant flavors, making the enzymes that may have dual functions in the process of developing flavor.
- **Other Carbohydrate-Degrading Enzymes, Including Amylases:** While not as important as proteases and lipases in degrading seafood quality, carbohydrate-degrading enzymes such as amylases also have an impact. These break down glycogen and other carbohydrates and, therefore, the sweetness and texture of the product. For instance, in crustaceans like shrimps and crabs, this post-mortem glycogen breakdown will yield a sweeter taste, usually preferred.

3.4.2. Impact on Texture and Flavor

The major determinant of seafood texture is their activity. For example, muscle tissues will be softened by hydrolysis of structural proteins such as collagen and myofibrillar proteins by proteolytic enzymes. This softening may be positive, resulting in tenderness or flesh palatability at an initial stage but excessive proteolysis will result in a mushy consistency, a very undesirable condition indicative of spoilage (Fernandes, 2016). Temperature, pH, and the specific protease inhibitors in the seafood account for some of the major factors affecting the rate and degree of proteolysis.

Lipase activity, through hydrolysis, leads to the formation of free fatty acids, thus giving rancid off-flavors mainly in fatty fish. The free fatty acids by themselves might be volatile and give off a strong odor, thus reducing the sensory quality of seafood. It is also possible for the free fatty acids to undergo oxidation, resulting in secondary products that are aldehydes and ketones with disagreeable odors. This is particularly a problem with seafood products stored for some time or under less-than-optimum conditions, where lipid oxidation can proceed unchecked.

Nucleotide degradation, particularly ATP, affects seafood taste. During the first post-mortem stage, ATP breaks down into IMP, an enhancer for umami taste and subsequently adds to its succulent, full flavor profile that fresh seafood is renowned for. With increased autolysis, IMP breaks down into hypoxanthine, which gives off negative flavors, i.e., bitterness in food. It is the equilibrium between these compounds that can be the determining factor of whether seafood is fresh or compromised.

In seafood, the enzymatic breakdown of carbohydrates is not as important as that of proteins and lipids. Still, the enzymatic breakdown of carbohydrates can impact taste and texture. As an example, the breakdown of glycogen in mollusks may be associated with a sweet taste. On the other hand, a high degree of breakdown can be coupled with the loss of desirable textural properties, for instance, firmness, a very important characteristic in species like shrimps and scallops.

3.5. MICROBIAL SPOILAGE OF SEAFOOD

Microbial spoilage is caused by a few microorganisms, in particular, bacteria, yeasts, and molds, which break down organic matter to produce off-odors, off-flavors, slime, and discoloration, among others (Odeyemi et al., 2018). These constituents of the microbial community are affected by seafood species, source environment, and handling practices post-harvest (Figure 3.2).

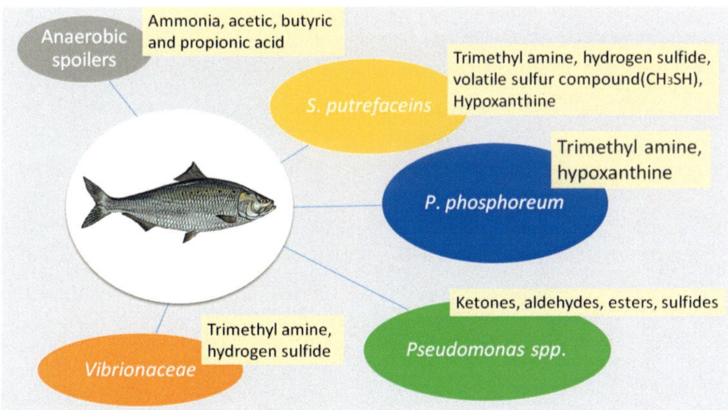

Figure 3.2. *Typical types of spoilage bacteria and typical compounds found during fish spoilage. Source: Salama, Youssef, Chennaoui, and Mohammed., 2024; Microbial Spoilage Organisms in Seafood Products: Pathogens and Quality Control. 1. 1. 10.5455/EJMID.20240518114533.*

- ***Pseudomonas spp.:*** Pseudomonas is a dominant spoilage bacterium in chilled, aerobic conditions. It is a psychotropic bacterium—what this means is that it can grow at low temperatures, making it particularly problematic in refrigerated seafood products. *Pseudomonas* spp. are known to be major producers of volatile sulfur compounds, among which hydrogen sulfide and dimethyl sulfide dominate, conferring the "fishy" smell to seafood during its spoilage process. These organisms are also able to synthesize extracellular enzymes able to degrade proteins and lipids, thus affecting the texture and flavor.

- ***Shewanella putrefaciens*.:** Another psychotropic bacterium, *Shewanella putrefaciens*, is one of the most abundant species found in marine environments and is considered to be a major spoilage organism in seafood. It produces TMA from TMAO, a molecule naturally occurring in many marine fish and responsible for that "fishy" smell often present in spoiled seafood. *Shewanella putrefaciens* also produces hydrogen sulfide and other sulfur compounds that give off off-odors.
- ***Vibrio spp.***: While a few species of *Vibrio* are pathogenic and cause foodborne illness, the majority are mostly involved in spoilage. *Vibrio* spp. can grow in seafood, specifically under warmer conditions. They can then produce off-odors and slime, particularly in shellfish like oysters and clams. The presence of these Vibrios is of concern not only for spoilage but also for food safety since certain of these strains can cause gastrointestinal illness.
- **Lactic Acid Bacteria (LAB):** Although primarily used in the process of fermentation, LAB can also be a cause of spoilage in vacuum-packed seafood. In such a process, these bacteria multiply in low-oxygen conditions to form lactic acid, which gives off sour odors and tastes. The incidence of LAB can alter the pH of seafood and hence affect other microbial growth.
- **Yeasts and Molds:** Less frequently identified than bacteria in seafood, nevertheless, yeasts and molds can be a source of spoilage in high salt or low pH products. Off-flavors, odors, and discoloration can be produced by these fungi, and in some cases, mycotoxins also occur. Visible spoilage of smoked or dried fish products occurs through the growth of some molds on the surface.

3.5.1. Factors Responsible for Microbial Growth

Temperature is one of the most important factors affecting microbial growth. As a matter of fact, most of the bacteria causing spoilage, particularly the psychotropic members like Pseudomonas and Shewanella, grow best at refrigerated temperatures between 0 and 10°C. At higher temperatures, however, spoilage becomes dominated by mesophilic bacteria, particularly by Vibrio, causing faster spoilage (Baptista, Horita, and Sant'Ana, 2020). Because of this fact, good temperature management through rapid cooling and maintenance of the cold chain is very important to slow down microbial activity, further extending the shelf life of seafood.

Availability or non-availability of oxygen can drastically affect which types of microorganisms may dominate the spoilage process. Whereas aerobic bacteria, such as Pseudomonas, would prefer an environment in the presence of oxygen, those kinds that are anaerobic or facultative anaerobic—like some Vibrio species and LAB may grow in low or no oxygen. Such low-O2 packaging methods as vacuum packaging or modified atmosphere packaging could, therefore, reduce the growth of aerobic spoilage organisms but favor the growth of anaerobes.

The pH of the seafood may impact the activity or growth of spoilage microorganisms. The optimum pH for growth of most bacteria is around neutrality or slightly alkaline, that is, from pH 6.5–7.5. However, some spoilage organisms, like LAB, may prefer more acidic conditions. Production of lactic acid by LAB will reduce the pH, making conditions that are inhibitory to some spoilage bacteria but may encourage the growth of yeasts.

Seafoods are also rich in nutrients, including proteins, fats, and minerals, all of which can sustain the growth of a varied microflora. Furthermore, seafood's high moisture content will also favor microbial proliferation. The initial microbial load on the seafood can make a big difference in the rate of spoilage.

Seafood has a high water activity, which acts on microbial development. The majority of bacteria require a high a w for growth, above 0.90, although yeasts and molds are more tolerant of lower a w levels. Given that the majority of preservation techniques reduce the aw by either drying or salting, such methods may prevent the growth of bacteria of interest and extend shelf life.

3.6. LIPID OXIDATION IN SEAFOOD

Lipid oxidation has adverse effects on the outcome of quality, nutritional value, and safety of seafood. It is the process of oxidative degradation mainly of lipids like unsaturated fatty acids, leading to the rancidity of seafood, the formation of off-flavors and odors, along with the formation of some potentially harmful compounds. Knowledge of factors affecting lipid oxidation applied with efficient techniques of detection and measurement would be vital for the maintenance and quality of seafood products.

3.6.1. Factors Influencing Lipid Oxidation

3.6.1.1. Degree of Unsaturation

Generally, the unsaturation degree of fatty acids has been cited as the main factor determining susceptibility to lipid oxidation. ω-3 Polyunsaturated fatty acids are abundantly found in fish and other aquatic animals and therefore need special consideration while preparing or processing. Polyunsaturated fatty acids, particularly eicosapentaenoic acid (EPA) and docosahexaenoic acid (DHA) from lipids of fish, have several double bonds and are readily oxidized. The rule is that the higher the degree of unsaturation, the more susceptible the lipid is to oxidation. Therefore, fatty fish, such as salmon, mackerel, and sardines, are more susceptible to oxidation than lean fish, such as cod.

3.6.1.2. Presence of Pro-Oxidants

While some enzymes like metal ions such as iron and copper, heme proteins, and certain enzymes may catalyze the oxidative process, in turn, metal ions could catalyze the oxidation of lipids by increasing the amount of reactive oxygen species (Undeland, 2016). Being iron-containing, heme proteins can catalyze lipid oxidation; consequently, lipid oxidation can take place in the muscle tissues of seafood. In addition, in seafood, lipoxygenases, a group of enzymes, can directly oxidize unsaturated fatty acids, thus leading to an increase in lipid oxidation.

3.6.1.3. Antioxidants

Lipid oxidation can be significantly affected by the presence of natural or added antioxidants. Antioxidants like vitamin E in the form of tocopherols, vitamin C in the form of ascorbic acid, and phenolic compounds can quench free radicals and decompose peroxides, and therefore inhibit lipid oxidation. Seafoods contain some natural antioxidants, but their levels are poor for oxidation to be retarded in oxidative storage conditions. It is for this reason that additional antioxidants incorporated into seafood products can help retard oxidation.

3.6.1.4. Environmental Conditions

Oxidation of lipids is highly affected by the presence of oxygen, light, and temperature, and these are mostly looked at as some of the key environmental factors. Exposure to oxygen will overly correspond to the rate of oxidation, while light—especially ultraviolet light—has a photolytic effect that will free radicals by induced oxidation. Temperature pushes the process further by increasing kinetic energy between molecules with increasing temperatures, thus catalyzing the oxidation reaction. Thus, appropriate packaging and handling are required to control these environmental factors and prevent or delay lipid oxidation.

3.6.1.5. Other Processing and Storage Factors

Processing and storage methods applied to seafood products can affect the extent of lipid oxidation. For example, mechanical processing is known to damage cell membranes, which release pro-oxidants and elevate the exposure of lipids to oxygen. Equally, storage conditions include temperature and packaging atmosphere. Oxidation can be considerably avoided or reduced by refrigeration and freezing, both through reduction of reaction rate and limiting oxygen availability. Modified atmosphere packaging (MAP) causes a reduction of the activity of oxygen due to the replacement of the oxygen with inert gases available such as nitrogen and carbon dioxide.

3.6.2. Oxidation Detection and Measurement Techniques

There are a number of techniques used to determine and gauge the level of oxidation in seafood products. Some of the methods include:

- Peroxide value measures the concentration of peroxides and hydroperoxides, major primary oxidation products. This is a widely used method for assessing the initial stages of lipid oxidation. The peroxide value is obtained by titrating the sample against a reducing agent, usually potassium iodide, and measuring the liberated iodine. A higher PV reflects a greater degree of primary oxidation. This method, however, only gives information regarding the early stages of oxidation and is not a measure of the secondary oxidation products.
- The Thiobarbituric Acid Reactive Substances (TBARS) assay is a test measuring malondialdehyde, a secondary oxidation product arising from the breakdown of polyunsaturated fatty acids. In the assay, MDA reacts with thiobarbituric acid to form a colored complex, which is then measured spectrophotometrically. In this respect, TBARS is one of the most frequently used methods, as it is not difficult to perform and is relatively sensitive, but it has a disadvantage: many compounds reacting with TBA would overestimate the values.
- The anisidine value indicates aldehydes, mainly 2-alkenals and 2,4-dienals, that are secondary oxidation products. AV is the absorbance of the complex formed by the reaction of the sample with p-anisidine. In this technique, information is provided on the secondary products of lipid oxidation and hence supplements the peroxide value in a manner that a more complete picture may be obtained for lipid oxidation.
- Gas chromatography is a method that is both very sensitive and specific for volatile oxidation products, such as aldehydes, ketones, and alcohols. Individual oxidation products can be separated and

quantified by GC, providing knowledge regarding the degree and nature of lipid oxidation. The technique is highly useful for the identification of some compounds responsible for off-flavor and odor formation in oxidized seafood.

- Fourier Transform Infrared (FTIR) spectroscopy can detect changes in the chemical structure of lipids that result from oxidation. In this method, it can measure the absorption of infrared light by various functional groups of lipid molecules, providing information on the formation of hydroperoxides, aldehydes, and other oxidation products. Since it is a non-destructive technique, FTIR can give quick analysis and therefore is suitable for routine monitoring of lipid oxidation.
- Nuclear Magnetic Resonance (NMR) spectroscopy provides detailed information on the molecular structure of lipids and their oxidation products. This technique identifies and quantifies many oxidation products that can give information on mechanisms and pathways of lipid oxidation. Nevertheless, despite being very informative, NMR is a rather complex technique that requires special equipment and expertise.

3.7. PROTEIN DENATURATION IN SEAFOOD

Protein denaturation is a process in which the structures of proteins are modified without breaking their peptide bonds, hence changing the physical, chemical, and functional properties of proteins. Several factors can induce protein denaturation, including temperature, pH changes, chemical agents, and mechanical stress.

Among the major factors that result in denaturation of seafood proteins is temperature. The heat during boiling, grilling, frying, or steaming leads to protein unfolding and consequently coagulation. Denaturation is important in food safety as it inactivates the pathogen and enzymes, that are responsible for spoilage. If there is too much heat, the proteins over-denaturate, lose moisture, and become tough and dry. On the other end, low temperatures, particularly during freezing, can also denature proteins. Freezing induces the formation of ice crystals, which may further perturb protein structures—the so-called "freeze-denaturation," is a process that influences the texture and water-holding capacity of seafood.

Another very important cause of protein denaturation involves changes in pH values. pH levels of seafood can either raise or lower due to metabolic by-products produced while in storage or from acidic or alkaline marinades. Each of the proteins has a normal characteristic pH range, described as the isoelectric point, where they are least soluble and most easily precipitated. If the pH is too far above or below the isoelectric point, they become denatured, leading to changes in their solubility and functional properties. For example, when lactic acid is formed and pH decreases in seafood, proteins coagulate, which affects the texture of the product.

Chemical agents contribute to denaturation. For instance, during seafood processing, bringing in salt may cause protein denaturation and loss of water, which affects both the texture and flavor. Chemical preservatives, very useful for extending the shelf life, can inactivate enzymatic action but with some negative effects on the protein structure. Sulfites, added as preservatives, might break disulfide bonds, leading to protein denaturation and hence affecting the quality of the product. Besides, mechanical stress given during various procedures such as filleting, mincing, and grinding can cause denaturation. Relevant in seafood products is the surimi process where fish proteins are extracted and refined by mechanical processing. Mechanical actions can disturb the native structure of the proteins, which may cause alteration in texture and functionality.

The denaturation of proteins in seafood is important in terms of nutritional quality. On the other hand, the denaturation process may make the proteins more digestible due to the exposure of peptide bonds to the digesting enzymes, but it may cause a loss of essential nutrients. For example, denaturation caused by heat can destroy heat-sensitive amino acids such as lysine, hence reducing the nutritional value of the protein. In addition, denaturation processes may be linked to the availability of other nutrients in organisms. For instance, in denatured proteins, water-holding capacity is lost, leading to the leaching of water-soluble vitamins and minerals and causing a diminution of the general nutritional profile of seafood.

The state of the protein in seafood is related to its sensory features, such as taste and texture. Well-denatured proteins contribute to palatability through favorable textures and flavor profiles. Too much denaturation leads to off-flavors and the loss of the naturally delicate texture, therefore, deviating from consumer acceptability. This would then be vital for seafood products with high quality, where texture and flavor are crucial qualities (Figure 3.3).

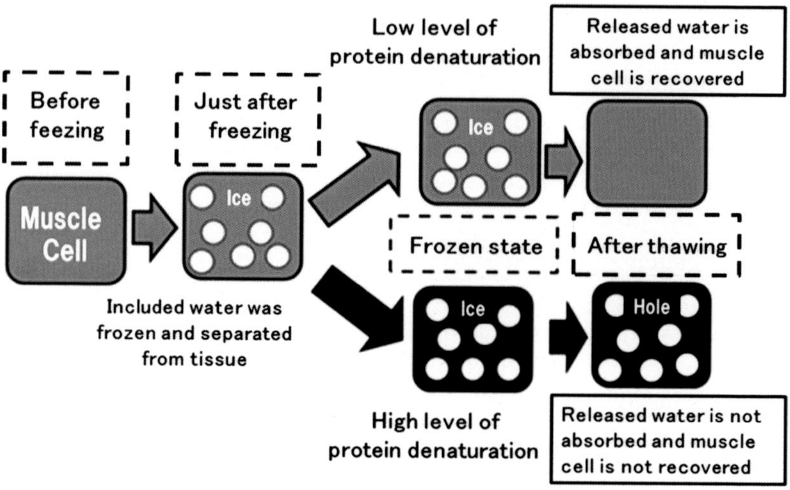

Figure 3.3. *Behavior of water and proteins in the process from freezing to thawing of muscle tissue of fish.*

Source: Rumape et al., 2022; Food Sci. Technol 42 https://doi.org/10.1590/fst.78521.

3.8. CHANGES IN CARBOHYDRATES IN SEAFOOD

The main types of carbohydrates found in seafood are glycogen, chitin, and different polysaccharides, depending on the species. These carbohydrates undergo several degradation processes during postharvest processing that eventually affect taste, texture, and hence the quality of seafood.

3.8.1. Carbohydrate Degradation Processes

In the majority of seafood species, and particularly in mollusks, crustaceans, and some fish, the major carbohydrate is glycogen. Glycogen represents a reserve of energy, which is stored in muscle tissues. The process of glycogen breakdown to glucose and other simpler forms of sugar begins shortly after catching, as a result of enzymatic activity. This process is known as glycolysis. In live seafood, glycogen is broken down to meet the metabolic demands of the creature. However, following death, this process will continue with the activity of endogenous enzymes like glycogen phosphorylase, even though the organism can no longer regulate the process.

During glycogen breakdown, lactic acid is formed, which can contribute to a decrease in the pH of seafood. This acidification will then modulate other enzymes' activities, causing the fast breakdown of another macromolecule of proteins and lipids. Lactic acid may cause a sour taste in seafood meat when glycolysis has taken place inappropriately at an increased or longer time due to poor storage conditions.

Another carbohydrate-related important constituent in fishery products, mainly in crustaceans, is chitin. It is a structural polysaccharide that forms part of the exoskeleton of crustaceans, including shrimps, crabs, and lobsters. While chitin itself is rather stable during postharvest processing, it can degrade into chitosan and other derivatives under certain conditions. Such degradation is usually brought about by chemical means rather than by enzymatic activities in seafood. In some algae and marine plants, the content includes such polysaccharides as agar, alginates, and carrageenin's. Such polysaccharides are generally not hydrolyzed by endogenous seafood enzymes but may be acted upon by exogenous enzymatic or chemical treatment. They are frequently extracted for use in foodstuffs as gelling agents.

3.8.2. Impact on Taste and Texture

Carbohydrate degradation in seafood may result in changes to flavor and texture. Glycogen may break down into glucose and other simpler sugars, which would give off a sweet taste in some fish, especially in the more carbohydrate-rich shellfish like oysters, scallops, and crabs. This has often been considered an attractive quality of these products, which adds to their palatability.

However, this will alter the taste profile should glucose be further converted to lactic acid, which will then introduce sourness or tartness. While a slightly acidic taste may sometimes improve the flavor profile, extreme acidification can lead to the development of off-flavors that are an indication of spoilage. Further, the depression of pH from the lactic acid formation shall also impact the texture of the seafood. Lowering the pH will denature proteins and make them coagulate, with associated textural changes in firmness and water-holding capacity of the tissue; this can then result in a mushy or very soft texture, which is undesirable and very perceptible in fish fillets and shellfish meat.

Chitin and its derivatives are odorless; however, they play a very critical role in the texture of crustacean seafood. The integrity of the exoskeleton is supported by chitin, which provides firmness or even crunch to shrimp, crabs, and lobsters. Any breakdown or weakening of the chitin can hence result in the softening of the shell, which could impact the apparent freshness and quality of such seafood.

Seaweeds, agar, alginate, and carrageenan's found in marine plants have a bearing on product texture and mouthfeel. Such chemicals are characterized by their gelling and thickening properties, which may lead to variation in texture when used as an ingredient in seafood dishes or processed products. The content and modification of carbohydrates can also become very important in products like surimi or fish paste. Added carbohydrates, mostly starch, may improve the texture and water retention of these processed seafood products. The enzymatic hydrolysis or chemical changes that these added carbohydrates undergo can then affect the texture of the final product to become either firmer or more elastic, depending on the preferred quality.

3.9. FACTORS AFFECTING BIOCHEMICAL CHANGES IN SEAFOOD

Biochemical changes that occur after seafood harvesting have immense effects on the quality and shelf life of seafood. These changes are influenced by temperature conditions, handling, storage, and processing methods.

3.9.1. Temperature and Storage Conditions

Temperature is probably one of the most important factors that affect biochemical changes in seafood. An increase in temperature raises both the rate of enzymatic activity and microbial growth; hence, seafood must always be kept at low temperatures to give it the longest possible shelf life. The main biochemical processes affected by temperature are autolysis, lipid oxidation, and protein denaturation.

Refrigeration considerably delays autolysis i.e., the self-digestion of the tissues by some endogenous enzymes. Some proteases, lipases, and nucleases, for instance, exhibit optimum activity at higher temperatures, hydrolyzing proteins, fats, and nucleic acids. This has been associated with the softening of tissue, off-flavors, and the formation of undesirable compounds. Provided the temperature is stored as close to 0°C as possible, these enzymatic activities can be significantly retarded, ensuring that texture and flavor are maintained in the product. Freezing seafood at temperatures below -18°C is more effective, because it practically suspends enzymatic reactions and microbial growth. It should be noted that freezing may cause protein denaturation and changes in the texture caused by the formation of ice crystals, which damage cell structures.

Another temperature-dependent process is lipid oxidation. Unsaturated fatty acids, when exposed to higher temperatures, develop more into rancidity, leading to off-flavors, off-odors, and harmful compounds. It will be further retarded by cold storage and freezing. Other practices related to packaging, those that reduce oxygen exposure, are vacuum packaging and modified atmosphere packaging, which are quite good at slowing down the oxidative processes.

Other conditions of storage that affect biochemical stability include humidity and exposure to light. High humidity will support the growth of microorganisms and cause spoilage, and light, particularly UV light, could act to catalyze oxidation reactions. In that consideration, seafood portions should be stored under conditions that are as dark, cool, and low in humidity as possible to maintain quality.

3.9.2. Methods of Handling and Processing

Methods of handling and processing significantly impact the biochemical stability and quality of seafood. Good handling must, therefore, begin right from the time when seafood is caught since rough handling causes damaged tissues and these lead to the released enzymes, which act to accelerate spoilage. For example, tissue damage may result in muscle cell rupture with released enzymes, that later provoke fast degradation of proteins and lipids.

Processing methods such as filleting, gutting, and cutting also affect biochemical changes. These expose more surface area to oxygen, hence increasing the possibility of oxidation reactions and microbial invasion. The mentioned operations need to be performed in a sanitary condition and as fast as possible to prevent exposure to degrading factors. Applying antioxidants directly or including them in packaging materials may help in depressing oxidation during processing.

Thermal processing through cooking, smoking, pasteurization, or any other means has the potential to denature proteins, inactivate enzymes, and kill off microorganisms of spoilage, thus improving shelf life. However, such methods may lead to losses in nutritional value and cause changes to the texture and flavor of the food. The application of heat should be well controlled to achieve a balance between safety and quality since excessive heat will form undesirable compounds and cause the loss of volatile flavors.

Other methods that do not require heating include HPP and irradiation; the latter methods can inactivate the enzyme and microorganisms without some of the adverse effects of high temperature. For example, HPP can kill bacteria efficiently and cause extended shelf life without changing sensory and nutritional properties in seafood. On the other hand, such methods usually present limitations for wide use because of the high cost and technical complexity.

Packaging provides protection of seafood from environmental factors likely to increase biochemical changes. Modified atmosphere packaging and vacuum packaging limit the amount of available oxygen, delaying oxidation processes and the growth of aerobic microorganisms. Research studies on active packaging technologies incorporating antimicrobial agents or oxygen scavengers are in process to further extend the shelf life and safety of seafood products.

3.10. METHODS TO CONTROL BIOCHEMICAL CHANGES IN SEAFOOD

Temperature management is one of the most efficient techniques for controlling the biochemical changes in fish. Such basic techniques as refrigeration and freezing retard enzymatic activities and microbial growth in seafood. When the temperature is lowered, metabolic processes in seafood are considerably retarded, thus increasing shelf life. Refrigeration is usually kept near 0°C and is suitable for short-term storage, whereas freezing, below -18°C, is used for long-term preservation. However, freezing can cause some physical changes, such as the formation of ice crystals, which may cause actual physical damage to cellular structure and hence affect texture. To minimize the problems caused by these phenomena, rapid freezing techniques are often used, such as blast freezing or cryogenic freezing.

Another critical preservation technique is the control of oxygen exposure. This might catalyze the oxidation of lipids and introduce rancidity and off-flavors. One such method is vacuum packaging and modified atmosphere packaging. In vacuum packaging, a package is completely cleared of air; the resultant low level of oxygen slows down the oxidation process. In contrast, MAP replaces the air with a specific mix of gases, mostly high in carbon dioxide or nitrogen, to further hinder oxidation and the growth of micro-organisms. These techniques of packaging are very effective at maintaining the quality of high-fat seafood species, such as salmon and mackerel.

Salting and drying are traditional seafood preservation methods that lower the water activity, a very critical determinant of microbial growth. The addition of salt dehydrates microbial cells through osmotic pressure, preventing them from duplicating. Drying may be done by merely air-drying, smoking, or freeze-drying food; all of these techniques reduce the moisture content and so make less water available for biochemical reactions and microbial growth. While such techniques preserve seafood, they also give specific flavors to the products, such as salted cod or smoked salmon.

Chapter 4

Quality Assessment of Seafood

Contents

INTRODUCTION

Quality assessment for seafood in the seafood industry is the most important issue and pillar for the seafood industry. This is a range of techniques and measures to evaluate freshness, safety, nutritional value, and overall acceptability of fish products. The quality of seafood readily deteriorates, so maintaining its quality from catch to consumer is very vital to consumer safety and satisfaction (Oehlenschläger, 2013). This chapter deals with seafood quality assessment techniques, including sensory evaluation, chemical analysis, microbiological testing, and instrumental methods.

Sensory evaluation is an important index to establish freshness and therefore desirability only using the human senses to evaluate any changes in appearance, texture, flavor, and aroma. This qualitative assessment gives instant feedback regarding the overall quality and therefore the acceptability of seafood products to producers, retailers, and consumers.

Chemical analysis is a more objective approach in fish quality control. Measured are certain chemical constituents that could indicate freshness, spoilage, or even contamination. Commonly applied chemical tests include those for lipid oxidation, protein degradation, and harmful substances such as heavy metals and biotoxins. Results from such tests are crucial to ascertaining regulatory compliance of seafood products and safety for consumption by consumers.

Another critical constituent in seafood quality assessment is microbiological analysis, which takes the form of detection and enumeration of pathogenic and spoilage microorganisms. This analysis ensures that seafood products are free from harmful bacteria, viruses, and parasites that may be a serious health hazard for consumers.

These instrumental methods utilize the technologies and instruments developed to improve seafood quality assessment capabilities. They provide explicit information on physical and chemical parameters regarding seafood, including texture, color, and composition, usually with a high degree of precision and accuracy.

Factors influencing seafood quality perception are vast, ranging from environmental conditions to methods of processing and storage, and even packaging materials. Regulatory bodies, on their part, set guidelines and standards that ensure seafood products are safe, of quality, and properly labeled. Similarly, consumer perceptions and preferences exert huge effects on the seafood market since perception and preference influence purchase decisions and brand loyalty. Emerging technologies can offer new opportunities for nondestructive and fast-quality assessments, making them highly sought after.

4.1. SENSORY EVALUATION OF SEAFOOD

Sensory evaluation in the assessment of seafood quality is prominent because it relates to the sensory attributes that consumers experience: appearance, odor, flavor, and texture. The human senses judge the quality and acceptability of the seafood under test and provide information one would not get through a solely conducted chemical or microbiological test. The process of sensory evaluation should, therefore, be conducted by predefined criteria and protocols so that the results can be consistent and reliable (Figure 4.1).

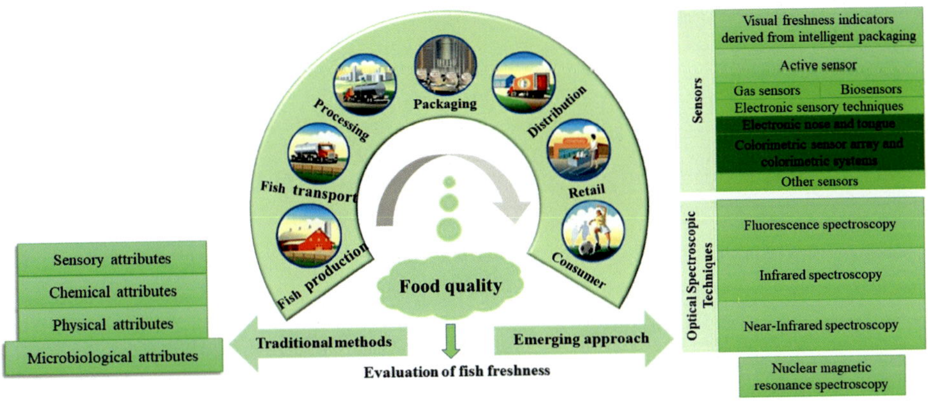

Figure 4.1. *Sensory evaluation on a fish.*

Source: Zhang Z, Sun Y, Sang S, Jia L, Ou C. Emerging Approach for Fish Freshness Evaluation: Principle, Application and Challenges. Foods. *2022; 11(13):1897. https://doi.org/10.3390/foods11131897).*

4.1.1. Sensory Attributes of Seafood

The sensory characteristics of seafood are very diverse and cover a number of important aspects. First to be noticed is the appearance: color, brightness, and general visual appearance (Çankırılıgil and Berik, 2018). Fresh seafood would typically come with very bright, vibrant colors and a moist, glossy appearance. Fish eyes should be clear and bulging, while skin should be bright and metallic. Changes in these visual cues may indicate deterioration or spoilage.

Odor is another very important sensory attribute. Fresh seafood would have an odor much like the sea, a description frequently given as clean and slightly briny. Any departure from this, such as sour or ammonia-like or sulfurous odors, may be characteristic of spoilage or contamination. This indicator is very sensitive since the majority of organisms causing spoilage produce volatile compounds that could be easily detectable even at quite low concentrations.

Flavor is similar to odor, but refers to the taste sensations when seafood is consumed. Fresh fish and shellfish have a delicate and subtle flavor but flavor differs greatly between species. Off-flavors, such as bitterness or sourness, indicate spoilage, the presence of biogenic amines, or contamination with environmental pollutants.

Texture is yet another essential sensory attribute of fishery products, more especially fish fillets, shrimp, and shellfish. The texture of fresh seafood should be firm and resilient. For example, fish fillets must feel firm to the touch, and the flesh should spring back when pressed. Mushiness, sliminess, or excessive dryness are indicators of spoilage.

4.1.2. Methods of Sensory Evaluation

The systematic methods for seafood sensory evaluation are designed to be objective and reproducible. These methods can be broadly categorized into descriptive analyses, difference tests, and affective tests.

Descriptive analysis refers to the process of in-depth assessment of predefined sensory attributes. The panel members would use standardized scales and descriptors to quantify a set of characteristics, including color intensity, odor strength, and firmness of texture (Kemp, Ng, Hollowood, and Hort, 2018). In the sense of providing a complete sensory description of seafood, this method could be very useful for quality control and product development.

Difference tests detect whether there are perceptible differences or not between two or more samples. The tests include a triangle test, duo-trio test, and paired comparison test. As an example, in the triangle test, the panelists will be presented with three samples; two of them will be alike, while one will be different. Panelists are then asked to select the odd sample. Such tests are pretty useful in the identification of minute changes in quality related to processing or storage conditions.

Affective tests measure consumer preference and acceptability. Untrained consumers who represent the target market evaluate the samples for overall liking and specific attributes. These could be very useful in giving insights into consumer perception and preference and thus guide product formulation and marketing strategies. Sample preparation and presentation are important in influencing the accuracy and reliability of the evaluation. The samples should be treated with care to prevent contamination and damage to their sensory properties. For instance, seafood samples must be prepared and tested under standardized temperature and lighting conditions. The application of neutral-colored containers and evaluation rooms minimizes extraneous impacts on the sensory assessment.

Such consistent and accurate evaluation calls for panel training. This means familiarizing the panelists with the sensory attributes and the descriptors that express seafood, as well as the proper use of evaluation scales. The panelists should also be monitored and calibrated regularly to hold up the quality and reliability of the sensory assessments.

Evaluation in the sensory laboratory is conducted based on approved guidelines that allow for objectivity and consistency. According to the guidelines in the uploaded document, facilities for the sensory evaluation will have separate areas for the preparation and testing of samples to avoid cross-contamination and assure controlled conditions. Panelists must be adequately trained, and their performance regularly monitored for consistency and reliability (CXG_031e).

While the quality assessment of seafood always includes sensory evaluation, it is a direct way to find out about the sensory attributes that are influential in consumer acceptance. This pertains to an in-depth examination of the appearance, odor, flavor, and texture of a given sample by structured methods like descriptive analysis, difference tests, and affective tests. To get the right result, correct appraisals about the sensory aspects should be obtained from well-trained panelists, following well-structured standardized protocols. The approach helps the seafood industries maintain high-quality standards, hence meeting consumer expectations.

4.2. CHEMICAL ANALYSIS OF SEAFOOD

Chemical analysis in seafood quality assessment forms an integral part of the delivery of accurate and objective measures for many chemical constituents that relate to freshness, spoilage, contamination, and general nutritional value regarding seafood. Such analyses are very important in establishing whether seafood products conform to the set regulatory standards and are safe for human consumption. Several types of chemical tests are commonly used in the assessment of seafood quality, all of which deliver some information on various aspects of the composition and condition of a given product.

4.2.1. Common Chemical Tests

There are a number of commonly used test programs or rather tests to analyze and establish the chemical architecture of seafood.

- The Total Volatile Basic Nitrogen Test abbreviated as TVB-N is run to determine the level of nitrogenous compounds-ammonia, dimethylamine, and trimethylamine-which are byproducts of microbiological protein degradation in seafood. High levels of TVB-N provide a pointer to spoilage since these compounds significantly contribute to the off-odors developed during the decomposition process for seafood.

- Secondly, Thiobarbituric Acid Reactive Substances (TBARS) assay is one of the most common tests conducted on seafood to measure the extent of the lipid oxidation process, which may lead to rancidity and off-flavors. The secondary oxidation product, malondialdehyde, is measured by this assay, which is based on its reaction with thiobarbituric acid in a colored complex formation. The extent of lipid oxidation is proportional to the color developed, which is measured spectrophotometrically. High TBARS values are indicative of oxidative deterioration, an undesirable feature in fatty fish.
- Third is Histamine Test. The levels of histamine are determined to assess the presence of biogenic amines, which can accumulate in seafood, more specifically scombroid fish like tuna and mackerel, due to bacterial decarboxylation of histidine (Nevado et al., 2023). High levels of histamine are responsible for scombroid poisoning, which involves rash, headache, and gastrointestinal discomfort. Histamine testing is therefore very important to assure seafood safety and compliance with the prescribed regulatory limits.
- Moisture content is also one of the important quality indices of fishery products, which relates to texture, weight, and shelf life. It is usually measured by drying a known weight of a sample and determining the loss in weight. The test determines the water-holding capacity, which is influenced by conditions of processing and storage of seafood.
- pH measurement is also widely applicable. pH is an indication of freshness and spoilage in seafood. Fresh seafood will always show near-neutral pH, though with spoilage, it can drop due to the formation of organic acids by microorganisms. In this respect, monitoring changes in pH can help in detecting spoilage and ensure that the seafood falls within the safe consumption limits.
- Mercury and Heavy Metal Testing is also used. These include heavy metals such as mercury, lead, and cadmium, which bio-accumulate in seafood and may present important health hazards. These metals are analyzed chemically using techniques like Atomic absorption spectrometry (AAS) or Inductively coupled plasma mass spectrometry (ICP-MS). Such tests are important in terms of establishing whether seafood meets prescribed safety standards and does not contain levels of heavy metals above regulatory limits.

- Lastly is Protein and lipid content analysis. It is necessary to establish the protein and lipid content of seafood for its proper nutritional labeling and quality control. While the Kjeldahl method still represents a widely applied procedure for determining protein content by measuring nitrogen content, methods of solvent extraction are often used in measuring lipid content.

4.2.2. Interpretation of Results

Upon testing, how do we get to interpret these results? The results obtained from the tests are then compared with standard values that are predetermined, which vary by region and often even by regulatory body. For instance, both the European Union and the United States fix specific limits for the levels of TVB-N and histamine in seafood that shall not be exceeded to ensure consumer safety. A TVB-N value above the threshold level indicates that the seafood is unfit for consumption due to spoilage, while high histamine levels suggest a risk factor for scombroid poisoning.

Interpretation of the TBARS values should be based on knowledge of the sensitivity of seafood species to lipid oxidation. For example, fatty fish are more prone to oxidative rancidity, and even moderate TBARS values may indicate significant quality deterioration. Similarly, moisture content and pH are interpreted concerning the expected fresh seafood range; deviations from these ranges may indicate a quality decline signifying either spoilage or poor handling practices.

The test results regarding mercury and heavy metal are then checked against the prescribed safety limits provided by health authorities such as the U.S. Food and Drug Administration and the World Health Organization. These limits bring with them a great health risk, especially for vulnerable populations like pregnant women and young children.

Analysis of the protein and lipid contents can determine nutritional labeling for seafood products. These two constituents should be well measured to give food labeling that would bring out nutritional benefits for the consumer concerning the consumption of high-quality protein and omega-3 fatty acids in seafood. These results help in quality control in terms of ensuring consistency in the composition of different batches of the products.

4.3. MICROBIOLOGICAL ANALYSIS OF SEAFOOD

Microbiological analysis deals with the detection and enumeration of pathogenic and spoilage microorganisms. The analysis ensures that seafood products are safe for consumption and in compliance with regulatory standards to safeguard public health. The presence of pathogenic microorganisms in seafood may lead to serious foodborne illness, while the presence of spoilage microorganisms will reduce sensory qualities and shelf life (Figure 4.2).

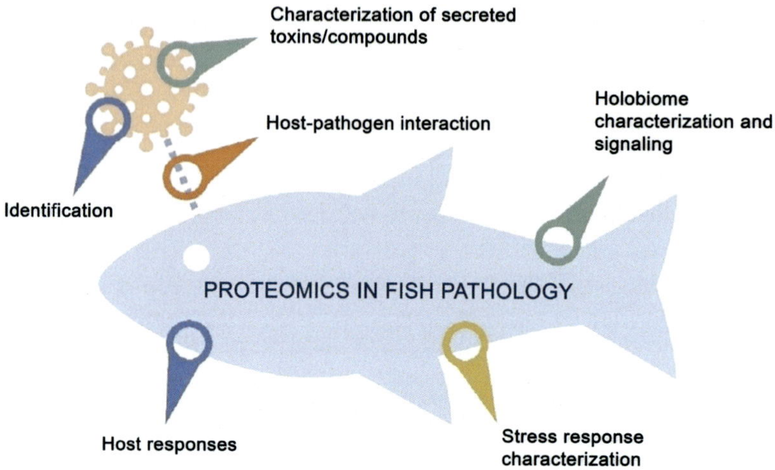

Figure 4.2. *Analyzing fish pathogens.*

Source: Moreira M, Schrama D, Farinha AP, Cerqueira M, Raposo de Magalhães C, Carrilho R, Rodrigues P. Fish Pathology Research and Diagnosis in Aquaculture of Farmed Fish; a Proteomics Perspective. Animals. 2021; 11(1):125. https://doi.org/10.3390/ani11010125.

Sections 4.3.1 and 4.3.2 will deal with the techniques that have to be used to detect pathogens in seafood and the importance of monitoring microbial quality.

4.3.1. Methods for Detecting Pathogens

Detection and quantification of seafood pathogenic microorganisms can be done with numerous methods, from traditional culture techniques to advanced molecular methods. All methods have their pros and cons; therefore, the choice of method is mostly based on the type of pathogen, the required sensitivity, and the resources.

- **Traditional Culture Techniques:** It is a classical method where isolation, detection, and enumeration of microorganisms are performed after their growth on selective media under specific culture conditions. Both of these are very commonly done because they are simple and useful in easily differentiating between different types of bacteria based on colony morphology and biochemical features. For instance, Salmonella spp. is isolated by pre-enrichment in nonselective media, then by enrichment in selective media, and finally plating on differential media such as Xylose Lysine Deoxycholate (XLD) agar. Also, Vibrio spp., usually through seafood, bacteria that can cause gastroenteritis, can also be detected by culture techniques. Although well-consolidated and reliable, these methods are very time-consuming and often need several days to deliver results.

- **Immunological Techniques:** Such as enzyme-linked immunosorbent assays ELISA and lateral flow assays, are based upon the detection of antibodies to specific pathogens or their toxins. They are relatively rapid and allow for large numbers of samples to be tested simultaneously. For instance, ELISA has been widely used for the detection of *Listeria monocytogenes* and *Escherichia coli* O157. Immunological methods are sensitive and specific, but potential cross-reactivity and false positives may warrant confirmation by culture techniques. Molecular methods can increasingly be found in the rapid and sensitive detection of seafood-borne pathogens, e.g., DNA or RNA of pathogens.
- **Polymerase Chain Reaction (PCR) and Quantitative PCR:** These techniques are now being increasingly used for the rapid and sensitive detection of pathogens in seafood. Since these methods detect the DNA or RNA of pathogens, even a low level of contamination can be identified (Zarei, Maktabi, and Ghorbanpour, 2012). PCR is particularly useful in the detection of pathogenic viruses, such as norovirus and hepatitis A virus, since these are non-culturable. qPCR not only confirms the presence of the pathogen but also gives information regarding the quantitative presence of the contamination, i.e., the level of contamination. Even though molecular methods are very sensitive and specific, they require special equipment and expertise.
- **Next-Generation Sequencing (NGS) Technologies:** They are rapidly getting established as strong tools with the capabilities of performing microbiological analyses that provide full profiling of microbial communities in seafood. NGS identifies a plethora of pathogenic bacteria, viruses, and parasites in just one test. The technology offers an in-depth look at the microbial composition of seafood and the possible risks associated with its consumption. The main limiting factors for the routine use of NGS in testing food safety are technology costs and complexity.

4.3.2. Importance of Microbial Quality

There are mainly two important reasons that microbial quality should be maintained by the seafood industry: assurance of food safety and prolongation of shelf life. Pathogenic microorganisms present in seafood can result in foodborne illnesses, which can vary from mild gastrointestinal to serious life-threatening illnesses. Some common seafood-associated pathogens are *Salmonella, Listeria monocytogenes, Vibrio* spp., *Escherichia coli*, and norovirus; each of these pathogens possesses different kinds of risks that require special handling and processing methods for the prevention of contamination.

For example, *Listeria monocytogenes* can proliferate at refrigeration temperatures, and thus it is the significant concern for ready-to-eat fish and seafood products. *Vibrio* spp. are common in warm coastal waters and may contaminate raw/undercooked seafood, mainly shellfish. Proper cooking and handling practices play a vital role in the removal of these pathogens to prevent outbreaks. Apart from these pathogens, there are also spoilage microorganisms that can cause a reduction in the quality and safety of seafood. The spoilage bacteria, such as *Pseudomonas* and *Shewanella*, may grow during storage and form off-odors, off-flavors, and slime. These signs of spoilage reduce the sensory appeal of seafood and indicate that it is no longer safe to consume.

Recall of seafood products on account of pathogenic contamination or evidence of spoilage can be very expensive to producers and retailers. Furthermore, incidents of foodborne illness or spoilage will greatly affect customer confidence in the seafood market, hence affecting the demand in the market and brand reputation.

In this regard, seafood producers should adhere strictly to hygiene and sanitation practices throughout the supply chain by placing the microbial quality in production from harvesting and processing to packaging and distribution. On a regular basis, there is a need to conduct microbiological testing so as to keep track of the microbial loading and detect contamination at an early stage of processing. Implementation of Hazard Analysis and Critical Control Points (HACCP) systems has assisted in identifying critical points where contamination can result and establishment of control measures for prevention or reduction of risks.

4.4. INSTRUMENTAL METHODS FOR QUALITY ASSESSMENT OF SEAFOOD

Assessment of seafood quality greatly relies on instrumental methods. These advantages include precision, objectivity, reproducibility, and accurate measurements. Instrumental methods are based on higher technological tools and instruments to measure various attributes of seafood, such as texture, color, composition, and freshness. The fact that instrumental methods can monitor and control seafood quality means that there are certain standards of products that should be yielded to consumers and regulators. The following sections consider some of the key technologies and instruments in use for seafood quality assessment, along with their applications and accuracy.

4.4.1. Technological Tools and Instruments

- **Spectrophotometry:** This is such an important technique applied to measure the color and composition of seafood products. The methods involve measuring the absorbance or reflectance of light by the sample at given wavelengths. These provide value in assessing pigment content, lipid oxidation, and protein concentration. For example, spectrophotometric methods may be applied when measuring the amount of heme pigments in fish. Such pigments provide the indication of freshness and color stability in fish.
- **Chromatography:** Gas and liquid chromatographic methods are very much necessary in investigating the chemical makeup of seafood. Foremost, gas chromatography is used for the separation and quantification of individual volatile compounds in seafood related to flavor and odor. Further, it is applied to determine compounds including pesticides and polychlorinated biphenyls (PCBs). On the other hand, non-volatile compounds including amino acids, vitamins, and biogenic amines are common to high-performance liquid chromatography (HPLC). All these techniques generate detailed profiles of the chemical constituents of a given plant material which stands very useful during quality control and safety evaluation.
- **MS (Mass Spectrometry):** Mass spectrometry is often linked with chromatographic techniques, such as GC-MS, and LC-MS. It represents a very sensitive, specific technique for the analysis of seafood chemical compounds. The technique identifies and quantifies a wide range of substances, from contaminants and toxins to nutritious components. The high sensitivity allows the detection at very low levels of compounds and is extremely useful in the assurance of seafood products' safety and quality.
- **Nuclear Magnetic Resonance (NMR):** These techniques provide comprehensive information regarding nuclear chemical environments and are also used in the molecular structure and composition of seafood. This technique can give much information on complex

mixtures of organic compounds and detailed determination of the lipid and protein composition in seafood.

- **Fourier Transform Infrared Spectroscopy (FTIR):** FTIR spectroscopy measures the sample's absorbance of infrared radiation, which provides information about its molecular composition. This technique is applied in fish quality assessment by detecting changes in molecular structures of proteins, lipids, and any other component in the seafood. More specific functional group changes caused by oxidation, denaturation, or spoilage can also be detected by FTIR.
- **Texture Analyzers:** These instruments measure the mechanical properties of seafood. They apply a controlled force to the sample and after that measure the response to acquire objective data on the textural attributes. Texture analysis is used in evaluating the quality of these seafood products, including fish fillets, shrimps, and shellfish, where texture reflects freshness and consumer acceptability.
- **Electronic Noses and Electronic Tongues:** The e-nose and e-tongue are composite sensing devices that respond to simulate the specificity of the human sense of smell and taste, respectively. E-noses consist of gas sensors that capture volatile compounds that give the sample its aroma profile; e-tongues consist of sensor arrays to capture taste-related compounds that give the sample its taste profile. These composite sensing devices are, therefore, very useful and competitive for seafood quality control by carrying out non-destructive assessments of the seafood for off-flavors and spoilage indicators.

4.4.2. Applications and Accuracy

Overdependence on a single instrument can be detrimental. Alternative instrumental methods should be considered for the assessment of quality in seafood and adapted to a set of applications with variable quality attributes and corresponding regulatory requisites: the accuracy of such methods is very important.

Spectrophotometry is hugely applied in color evaluation and pigment analysis, and it provides objective measures which greatly correspond with the visor. The spectral accuracy depends on several factors, including among others sample preparation and stability of the light source and calibration of the wavelength. Chromatography, with the coupling of mass spectrometry, offers a very good degree of accuracy and sensitivity for the detection of contaminants, nutritional components, and flavor compounds. These basic techniques are required for conformity to safety standards and nutritional labeling. NMR and FTIR spectroscopy provide very detailed molecular information, with high accuracy and reproducibility. They are indispensable in the study of protein and lipid composition and structural changes to support spoilage assessment and quality degradation. Using a texture analyzer measures mechanical properties with very high precision; this mostly is important for quality control studies in processing and packaging.

E-noses and E-tongues are employed for the rapid and non-destructive characterization of sensory attributes; thus, they are methods complementary to the traditional sensory assessment. The accuracy of these systems, on the other hand, depends to a large extent on the calibration and training of the sensor arrays and the consistency of presentation of the samples.

The basic accuracy of instrumental methods improves with rigorous calibration, standardized protocols, and procedures for regular equipment maintenance. These methods deliver objective data, which therefore justify underpinning quality assurance, regulatory compliance, and product development in the seafood industry.

4.5. FACTORS AFFECTING QUALITY ASSESSMENT OF SEAFOOD

4.5.1. Water Quality

This refers to the water quality in which it is harvested, which forms the bedrock of seafood quality in general. Some of the factors affecting water quality include temperature, salinity, pH, and the level of contamination. Poor water quality may lead to the concentration of harmful compounds such as heavy metals like mercury and lead or industrial contaminants like PCBs within the seafood, which can cause health problems when ingested. Moreover, the temperature of the water determines the metabolic rate of the fish, thus their fat content and flavor. Generally, colder waters will provide fish with a higher percentage of omega-3 fatty acids, which are heart-healthy. On the contrary, warmer water conditions raise the potential for pathogen contamination, such as Vibrio species that feast on higher temperatures.

4.5.2. Catching Methods

Various catching methods during the seafood collection process can significantly affect its quality. Catching methods like trawling or gillnetting are very likely to cause physical damage to fish, hence compromising their texture and appearance. Thereby, such damage would fasten the degradation process post-harvest since bruised or cut fish are more prone to microbial contamination and microbial spoilage. Line fishing and hand picking are some gentle and fast ways of harvesting that would minimize stress and damage to the seafood, thereby preserving its quality. Moreover, the time of the year and location of harvest may also influence the fat content, flavor, and texture of such seafood, as these are normally related to the reproductive cycles and feeding patterns of the organisms.

4.5.3. Post-Harvest Handling

The quality of seafood is maintained further through the process of post-harvest handling. This starts from the preliminary cooling of seafood to transportation and storage. This is important in reducing the enzymatic activity and microbial growth responsible for spoilage. Ice or chilled seawater, applied immediately after catching, helps keep the produce at low temperature levels, thus freshness is retained. In addition to that, the hygiene and sanitary state of handlers and equipment are very important in preventing cross-contamination. Poor handling practices, which involve exposure to high temperatures and dirty equipment, are capable of causing a dramatic decrease in both the shelf life and safety of seafood products.

4.5.4. Handling Techniques

A wide variety of handling techniques—filleting, deboning, and shelling, amongst others, are all to be carried out with due care to avoid damage to the flesh, thus affecting texture and appearance. Processing can also add or subtract some components to seafood. Added salt and smoking, for example, are excess elements in processing that enhance flavor and preservation but create high sodium content when used to excess, therefore affecting healthfulness. Cooking methods are very important as well, where steaming and poaching tend to leave the natural flavors and nutrients of seafood intact better than frying, adding many unhealthy fats and covering the delicate flavors of the seafood.

4.5.5. Storage Conditions

The length and quality of the shelf life are determined to a great extent by the conditions of storage. Temperature is a very important factor, and seafood should be stored always at a sufficiently low temperature to prevent spoilage and the growth of harmful bacteria. Freezing is a common mode of long-term storage, but it also demands adhering to the proper procedures to maintain optimum texture, to avoid freezer burns, and for proper packaging and handling techniques. Another critical aspect of storage is packaging; either vacuum packaging or modified atmosphere packaging can sometimes increase the shelf life of seafood dramatically by reducing oxygen exposure and thereby inhibiting microbial growth. Nevertheless, bad packaging or inconsistent temperature control can easily result in rapid spoilage off-flavors, and textural changes.

4.6. PACKAGING AND STORAGE OF SEAFOOD

Packaging and storage are very critical to the quality, safety, and shelf life of seafood. The choice of materials used for packaging and the conditions under which seafood is stored directly influence freshness, flavor, texture, and nutritional value. Proper dehandling, handling, storing, and processing techniques avoid spoilage, contamination, and quality deterioration at all stages of the chain.

4.6.1. Types of Packaging Materials

The different types of packaging materials used in the seafood industry differ in their properties and applications. Such major forms include plastic films, vacuum packaging, modified atmosphere packaging, and the more recent ones: biodegradable and sustainable ones.

- **Plastic Films:** They mainly polyethylene, polypropylene, and polyvinyl chloride, find applications in seafood packaging because they have good barrier properties with respect to water vapor and oxygen. They can either be transparent to allow the consumer to view the produce or opaque to protect products that undergo degradation in the presence of light. However, plastic films have limitations associated with environmental issues since most of them are non-biodegradable and contribute to plastic wastes.
- **Vacuum Packaging:** It is a process in which the air from the pack is sucked out before it has been completely sealed, thereby creating a vacuum condition. This mostly eliminates oxygen and prevents the action of aerobic bacteria, thereby arresting oxidation processes that often result in spoilage. The main application of this method is in the preservation of seafood products such as smoked fish, fillets, and shellfish. It also helps in preserving moisture, which retains the texture and juiciness of the seafood (Figure 4.3).

Figure 4.3. *Application of vacuum skin packaging to seafood products.*

Source: Kontominas MG, Badeka AV, Kosma IS, Nathanailides CI. Recent Developments in Seafood Packaging Technologies. Foods. 2021; 10(5):940. https://doi.org/10.3390/foods10050940.

- **Modified Atmosphere Packaging (MAP):** It involves a package technique where air is replaced with a mixture of gases, such as carbon dioxide, nitrogen, and oxygen. The gas composition in the mix is usually variable depending on the species of the seafood. For example, high carbon dioxide levels might prevent microbial growth, while high levels of nitrogen are used against oxidative rancidity in fat fish. The MAP process extends shelf life in seafood by maintaining the best gas conditions that promote freshness and prevent spoilage.
- **Biodegradable and Sustainable Packaging:** Environmental issues make the seafood industry fast embrace biodegradable and sustainable packaging materials. These include materials made from plant-based polymers like polylactic acid and from renewable resources like seaweed. Biodegradable packaging has properties almost similar to conventional plastics in protection but decomposes readily in the environment, hence reducing the ecological footprint.

4.6.2. Regulations and Standards for Seafood Quality

The quality and safety of seafood must be guided by an extended framework of regulations and standards from a variety of national and international regulatory authorities. These are to protect public health by making sure trading practices are fair and the integrity of seafood supply chains cannot be doubted. Compliance is assured through inspection, certification, and penalties in case of failure.

The simplest of these regulative bodies operating these markets at the most basic level are the United States Food and Drug Administration (USFDA), the European Food Safety Authority in the European Union, and at the international level, the Codex Alimentarius Commission providing standards for food. Controls should be put into place along with hazard analysis by the seafood industry under the enforcement of the Food Safety Modernization Act (FSMA) by the FDA. The FSMA was enacted to transform seafood safety from a reactive approach to a preventive one. This calls for the requirement of hazard analysis and HACCP systems that locate hazards within the processing line and take measures to control contamination.

At the EU level, EFSA sets high standards in relation to seafood safety regarding contaminants, additives, and labeling. The EFSA works in partnership with the European Commission and the member states to ensure that there are standard levels within the EU. The maximum residue limits include MRLs for contaminants, like heavy metals and veterinary drug residues, to make sure seafood products are safe to consume. The Codex Alimentarius Commission, in its joint venture with WHO and FAO, develops international food standards, guidelines, and codes of practice. These international standards stipulate a wide spectrum of concerns in foodstuffs, from hygiene to pesticide residues and finally to food labeling, providing a yardstick globally for the quality and safety of seafood.

Compliance with these regulations is very vital to seafood producers and exporters, as failure to comply may lead to severe penalties which include but are not limited to product recall, fines, and bans on exports. Compliance is normally ensured through a combination of internal and external audits, inspections, and testing. The processing premises of seafood products are inspected by regulatory bodies so that hygiene standards can be maintained with proper labeling and correct documentation. These can be scheduled or unannounced and thus introduce an element of unpredictability, creating an incentive for continuous compliance.

Certification schemes, such as the Marine Stewardship Council certification and Aquaculture Stewardship Council certification, provide buyers with additional assurance about the quality and sustainability of seafood. To be certified, producers have to meet strict standards regarding their impact on the environment, sustainable practices, and social accountability. This gives the certified products stronger consumer and retailer demand, thereby creating a market advantage for such producers who abide by the requirements.

Apart from these national and international regulations, there could be standards and requirements on top of that because of local conditions and concerns. For instance, Japan would have far stricter standards regarding the freshness and quality of seafood and may require more frequent testing for radioactive contaminants because of its unique environmental conditions. Similarly, countries with a large aquaculture sector may have particular legislation on the use of antibiotics and other veterinary drugs that reflect local practice and consumer concern.

The regulatory enforcement of seafood is very complex, with a lot of government agencies and private enterprises involved. Governmental inspection and audit are supplemented by third-party laboratory testing for an array of contaminants: heavy metals, biotoxins, and microbial pathogens. This testing is done to determine if products comply with safety standards and to inform consumers.

4.7. CONSUMER PERCEPTION OF SEAFOOD QUALITY

Quality perception by the consumer of seafood is very broad, including sensory properties, nutritional value, safety, sustainability, and labeling and branding efficiency. These play an important role in the formation of consumer preferences and, subsequentially, purchasing decisions.

This is important to facilitate seafood quality through the consumer's eyes. Sensory attributes of the product are among the most important factors in delivering seafood quality through the eyes of the consumer: appearance, taste, texture, and smell. Consumers tend to associate bright, vivid colors with freshness, and seafood with dull or off-colors may be perceived as not very fresh or of poor quality. Consumer acceptance is greatly affected by the texture of the seafood, whether it is firm or soft, and off-odors like ammonia or fishiness.

Another big concern of consumers is nutritional value. Seafood is rightly accepted as very healthy, more so in its content of omega-3 fatty acids, protein, vitamins, and minerals. As consumer awareness about the different nutrition contents of various seafood species increases at a very high rate, consumers might opt for products that could help in meeting specific health needs such as heart health, weight management, or brain function. It is such market awareness that fuels demand for species rich in omega-3s, like salmon, mackerel, and sardines.

Other major concerns include safety. Mercury, PCBs, and microplastics in tainted fish, as well as the impact of news reports and health advisories on the perception by the consumer. Such awareness may be enhanced by further problems like antibiotic use in aquaculture or the risk of foodborne diseases such as Listeriosis or Vibriosis.

Increasingly, consumers appear to be raising their awareness of sustainable and ethical issues. Indeed, advised about overfishing, bycatch, and the negative environmental impacts of aquaculture practice, more consumers are increasingly tipping towards sustainable seafood choices. Certifications by the Marine Stewardship Council or Aquaculture Stewardship Council, for example—are highly instrumental in giving consumers information on seafood sustainability. Sustainability today serves as a major trait in differentiating brand names within seafood products, more so with the emergence of the eco-consumer into the marketplace.

Labeling and branding influence consumer perception and choice. Good labeling will indicate the species name, country of origin, method of production, and certifications for sustainability or safety. If labeled clearly and correctly, the consumer can make good choices; some consumers are concerned about species substitution or mislabeling associated with the seafood sector.

Seafood product branding also influences consumer perception. Those brands touting quality, sustainability, and transparency win the trust and loyalty of consumers. For example, brands specifically advocating sustainability, social equity, or local sourcing will speak to environmentally or socially conscious consumers. In addition, such branding allows differentiation in a crowded marketplace, making it easier for consumers to identify premium or specialty products.

Attractive packaging enhances the perceived value or quality of the product; it shows or illustrates informative, engaging designs that teach customers about the benefits of seafood, the story behind the brand, or the environmental standards and values upheld by a company.

4.8. EMERGING TECHNOLOGIES IN SEAFOOD QUALITY ASSESSMENT

New technologies are changing seafood quality assessment into more efficient, effective, and non-invasive methods of assessment of the quality of the product. Improvements are important in ensuring food safety and increasing confidence among consumers in the seafood sector. The development of spectroscopy techniques, especially near-infrared spectroscopy and hyperspectral imaging, has been quite important. Such technologies support the rapid, noninvasive examination of seafood for moisture content, fat composition, and many other contaminants, all without sample preparation.

The application of biosensors and electronic noses is another important development in this area. The use of biological molecules makes biosensors highly selective and particularly sensitive to certain contaminants or pointers of quality. They are most helpful for *in vivo* monitoring of spoilage markers, mainly biogenic amines and volatile compounds. E-noses are designed to mimic the human olfactory system in detecting and differentiating complex odors; thus, they are quite fitting for freshness assessment and off-odors indicative of spoilage. These devices offer fast, objective, alternative methods to traditional techniques of sensory evaluation.

Digital imaging technologies, including 3-D imaging and computer vision, are also rapidly developing for seafood quality evaluation (Mathiassen et al., 2011). These methods can estimate quality indicators that can be visually perceived about seafood: color, texture, size – those things that provide a consistent and objective appraisal, especially in sorting and grading processes.

Apart from these technologies, blockchain and other computerized traces are being rolled out to allow greater traceability and transparency in seafood supply chains. All of these systems log a huge amount of information about seafood product origin, handling practices, and transport and hence prove the quality of food and no case of fraud exists. Blockchain technology can offer customers better and clearer information about the seafood they buy regarding certification on issues such as sustainability and the production method.

In the future, the application of artificial intelligence and machine learning algorithms in such advanced technologies should continue to yield very high accuracy and efficiency in seafood quality assessment. AI can analyze vast data sets for patterns that enable the prediction of quality outcomes, hence the putting in place of proactive control measures. This way, seafood supply management could ensure reduced wastage and that only products of the best quality reach the market.

Chapter **5**

Processing Technologies for Seafood

Contents

INTRODUCTION

Processing is an important factor in the industrial production of any given food. It is far more important in the seafood industry being critical for the preservation of quality, prolongation of shelf life, and assurance of safety from the point of catch or harvest to the consumer's table. Effective processing techniques are quite important in the processing of highly perishable seafood to prevent its spoilage, reduce wastage, and maintain the nutritional and sensory properties that make seafood one of the most valued parts of the human diet (Boziaris, 2014). From traditional methods passed down through generations to innovations mainly forced by the influence of technology and science, a good number of processing methods have been developed and refined over the centuries.

Techniques such as salting, smoking, curing, and drying are indeed ancient methods of processing seafood without the use of refrigeration, techniques that not only prolong the shelf life of seafood but also introduce various flavors and textures desired in many cuisines. While these technologies are very historic in nature, they are still quite relevant today, often combined with modern technologies to enhance their effectiveness and appeal.

In contrast, processing technologies have innovated the seafood industry in terms of more efficient, controlled, and scaled solutions to preserve and improve seafood quality. Innovations such as freezing, vacuum packaging, modified atmosphere packaging, and high-pressure processing have raised new standards to freshness and safety. These technologies help maintain the sensory and nutritional qualities of the seafood, reduce contamination risks, and meet the requirements needed for global distribution networks.

This interplay between traditional and modern techniques reflects the dynamism involved in seafood processing. As consumer preference changes with time, and new markets open up, there is a constant drive towards new and better ways of processing. Advances in packaging technologies have contributed immensely in extending the shelf life of seafood products and enhancing quality retention during transportation and storage, apart from offering consumer convenience.

Processing of seafood is done with high potential dangers of contamination from both microbial and chemical sources taken into consideration. Regulatory frameworks and guidelines on best practices are hence called for to ensure processing operations are as per the set standards and render products that are safe for consumption. It encompasses rigorous monitoring of contamination risks, strict adherence to hygiene procedures, and the setting in place of traceability systems aimed at tracing seafood from harvest to market.

5.1. TRADITIONAL METHODS OF SEAFOOD PROCESSING

All across the world, seafood has been and continues to be one of the biggest yet admired sources of food. Seafood dates back to ancient history marking a period where men and similarly women in the industry had to come up with ways of preserving and processing seafood products. These ways or rather, methods were developed out of necessity for preserving seafood where modern refrigeration did not exist and have been developed over the years to improve flavor, texture, and shelf life of seafood.

5.1.1. Historical Overview

The preservation of seafood has been around for 1000's of years, critical for communities that relied on fishing as a primary food source. Ancient civilizations, from the Egyptians and Romans to the Vikings and indigenous peoples of the Americas and Asia, developed preservation techniques in large numbers, ensuring a supply of fish was continually available throughout the year. These methods not only allowed excess catch to be stored during times of plenty, thereby freeing people to do other things but also ensured a reliable supply of protein in lean periods.

Salting is an ancient method of seafood preservation whereby salt is applied to fish or shellfish to draw out moisture through osmosis, making an environment very unfavorable to bacteria and other organisms of decay. Historical documents state that the Egyptians were already preserving fish in salt as far back as 2000 BCE. The Romans, in turn, developed complex methods of salting that gained highly valued products, such as garum—a fermented fish sauce. Salting not only lengthened seafood's shelf life but also made it more flavorful.

Another ancient process is smoking, whereby the seafood is subjected to smoke from burning wood. This does not only give a distinct flavor, smoky in nature but also preserves by way of reduction of moisture content and the introduction of antimicrobial compounds in smoke. Smoking salmon and trout is traditionally done to preserve fish by the indigenous people of North America, Scandinavia, and Siberia. Smoked products like those were crucial for survival through bitterly cold winters and long journeys.

The curing process, often combined with salting and smoking is another that has been used to produce a wealth of traditional seafood products. The curing process improves flavor and texture while increasing shelf life. Examples include gravlax, a Scandinavian dish made with salmon cured in a mixture of salt, sugar, and dill, and biltong, a type of dried and cured fish popular in Southern Africa.

Drying involves the removal of moisture from seafood to prevent spoilage and is, therefore, one of the easiest and oldest techniques of preservation. Sun-drying was used in regions that had favorable climates, while air-drying and wind-drying were common in cool areas. The economies of the coasts flourished during medieval Europe through the use of dried fish like stockfish and cod as essential trade goods, which in turn allowed long-distance trade.

5.1.2. Techniques and Their Significance

Each traditional seafood processing method is of value and significance in its own domain. They give value to the culinary heritage and contribute to the economic development of a region. Some examples of these impacts include:

- **Salting:** However ancient, remains very important and is one of the essential methods for processing and flavoring seafood. This technique is applied in the manufacture of salted fish products such as bacalhau of Portugal and Norway, which are cultural icons of food habits. They feature as main ingredients in traditional recipes in Portugal and Norway. It is similarly used in the processing of fermented seafood products, such as Japanese shiokara, a fermented fish viscera delicacy (Figure 5.1).

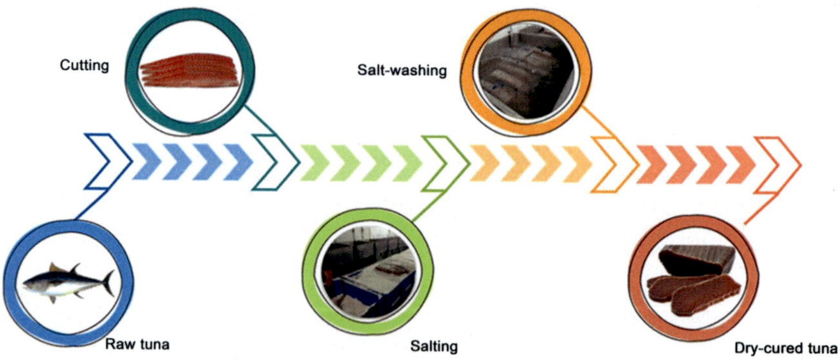

Figure 5.1. *Salting fish preservation method.*

Source: Sánchez-Parra M, Ordóñez-Díaz JL, Pérez-Aparicio J, Moreno-Rojas JM. Physicochemical and Microbiological Changes Associated with Processing in Dry-Cured Tuna. Applied Sciences. 2023; 13(10):5900. https://doi.org/10.3390/app13105900.

- **Smoking:** There is a special flavor profile imparted to seafood through smoking, making it a popular method of preservation for delicacies such as smoked salmon, kippers, which are smoked herring and smoked eel. It is important for the unique taste and texture it gives to the seafood than it is for preservation. These smoking techniques range from cold smoking, which occurs at lower temperatures to preserve delicate textures, to hot smoking, which actually cooks the fish while giving it a smoky flavor.

- **Curing:** This is a combination of processes: salting and smoking. Curing has already played a huge role in the production of a great variety of seafood products and is well on its way to becoming very significant in this field. Such treatments help evoke complicated tastes and textures so important in gastronomic traditions from all over the world. Products such as bottarga from Italy, mojama from Spain, and saltfish from the Caribbean contribute to staple ingredients in many local dishes.
- **Drying:** Although simple, drying has nevertheless been practiced for centuries as a means of preserving seafood. Dried fish products are part and parcel of traditional cuisines, either as flavoring or main course, such as Japan's katsuobushi, which is dried bonito flakes, and Korea's dried pollock. Furthermore, it is vital in terms of international trade, as dried fish are lightweight, non-perishable, easily transportable, and, as such, a very valuable commodity (Figure 5.2).

Figure 5.2. *Dried fish.*

Source: By Jean-Louis Vandevivère – flickr, CC BY-SA 2.0, https://commons.wikimedia.org/w/index.php?curid=13510937.

5.2. MODERN METHODS OF SEAFOOD PROCESSING

Modern processing technologies have changed the seafood industry in terms of efficiency, quality, and safety (Hassoun et al., 2022). The following examples of such modern seafood processing methods, which have completely revolutionized the industry.

5.2.1. Blast Freezing

Blast freezing is when the seafood products are rapidly frozen by powerful air currents that bring down the temperature. This will avoid the growth of large ice crystals, which usually cause damage to cellular structure and lead to loss of texture and flavor of the seafood.

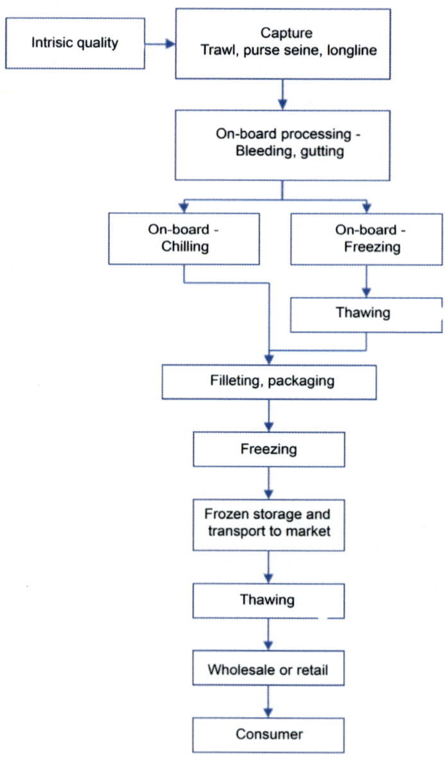

Figure 5.3. *The freeze-chilling value chain .*

Source: Erikson U, Uglem S, Greiff K. Freeze-Chilling of Whitefish: Effects of Capture, On-Board Processing, Freezing, Frozen Storage, Thawing, and Subsequent Chilled Storage—A Review. Foods. 2021; 10(11):2661. https://doi.org/10.3390/foods10112661.

Blast freezing is a very effective method and can be applied on a large scale; it is therefore widely used in freezing fish fillets, shrimps, and other kinds of seafood. The rapid freezing process also retains freshness and significantly increases shelf life, allowing seafood to be distributed and stored for a long period over vast distances without losing its quality (Figure 5.3).

5.2.2. Individually Quick Freezing (IQF)

Individually Quick Freezing (IQF), is the process wherein each piece of seafood is frozen singularly in very low temperatures. This technique does not allow clumping and allows consumers to use just the amount they need without defrosting the entire package. IQF is commonly used for products like shrimp, scallops, and fish fillets. This gives good convenience to the consumers, while at the same time decreasing food wastage. Quick freezing also allows maintenance of the nutritional value and sensory qualities of seafood, hence becoming an ideal choice for maintaining high-quality standards.

5.2.3. High-Pressure Processing (HPP)

High-Pressure Processing involves the use of extremely high pressure to inactivate microorganisms and enzymes responsible for deterioration. It is a non-thermal process. This technique does not require high temperatures compared to traditional methods, hence better preserving the nutritional and sensory qualities of seafood. This process works well on all raw and slightly processed seafood products, including even oysters and sushi-grade fish, which can easily be used directly from the processor without further cooking. HPP enhances food safety by eliminating pathogens while maintaining fresh taste and texture.

5.2.4. Modified Atmosphere Packaging (MAP)

Modified Atmosphere Packaging is a technique of in-pack manipulation of gases aimed at extending the fresh period of seafood (Stammen et al., 2009). It does this by inhibiting the growth of all kinds of spoilage microorganisms and oxidative reactions involved in the process due to reduced oxygen and increased carbon dioxide and nitrogen in the atmosphere. It finds an application domain in fresh and ready-to-consume seafood products. MAP is capable of maintaining the color, texture, and flavor of seafood and thus offers an attractive appearance to consumers. It is also known to reduce the need for chemical preservatives, hence a healthier option for packaging seafood.

5.2.5. Ultraviolet Irradiation

UV irradiation is the process of exposing seafood to UV light to reduce microbial contamination on its surface. More specifically, this method is useful in enhancing the safety of seafood mainly consumed raw, such as sushi and sashimi. UV irradiation is a non-chemical and residue-less approach that does not have any sensory effect on seafood. This significantly lowers the chances of foodborne illness by killing pathogenic bacteria and viruses present on the surface of seafood products.

5.2.6. Electromagnetic Heating

Application of electromagnetic heating includes microwave and radiofrequency heating to heat seafood products in a flash. Supply of power is excellent in uniformity for the reduction of overcooking and under-cooking conditions. Electromagnetic heating is efficient, hence saving a considerable amount of processing time. Such process is very useful for cooking and pasteurizing seafood products so that they are safe for consumption yet still retain as many nutritional and sensory qualities as possible.

5.2.7. Cryogenic Freezing

Cryogenic freezing uses ultra-low temperatures that typically consist of liquid nitrogen or carbon dioxide to freeze seafood at a very fast rate. As it contains fast freezing, there will be minimal formation of ice crystals, which can help retain the texture and quality of the seafood.

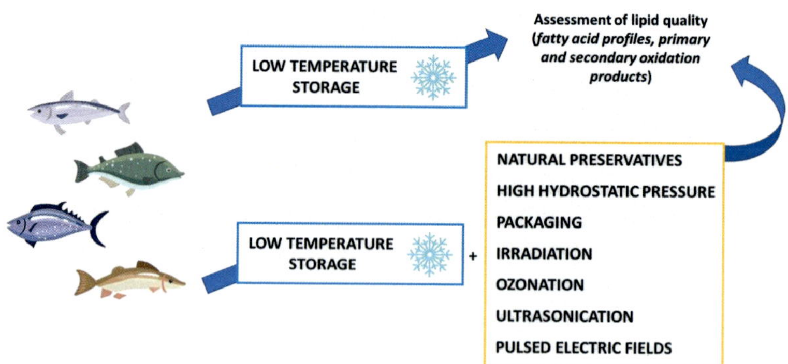

Figure 5.4. *Fish freezing.*

Source: Suárez-Medina MD, Sáez-Casado MI, Martínez-Moya T, Rincón-Cervera MÁ. The Effect of Low Temperature Storage on the Lipid Quality of Fish, Either Alone or Combined with Alternative Preservation Technologies. Foods. *2024; 13(7):1097. https://doi.org/10.3390/foods13071097.*

Effectiveness in cryogenic freezing lies in the freezing of delicate items of seafood such as fish fillets, shrimp, and shellfish. This method ensures superior quality in terms of preservation when compared to traditional techniques of freezing and finds application within the seafood industry to ensure standards of freshness and texture (Figure 5.4).

5.2.8. Benefits Over Traditional Methods

Advanced technologies such as HPP and MAP show increased control of microorganisms, which are contaminants and spoilers, reducing the risk of foodborne illness and improving seafood's shelf life. These techniques also allow for better retention of the nutritional and sensory qualities of seafood compared with traditional methods of processing, allowing for the flavor, texture, and nutritional content of the product to be retained.

Other key benefits include greater efficiency and the ability to scale up with the types of processing that are possible today. Process innovations like blast freezing and IQF allow producers to process seafood quickly and on a large scale; huge volumes of products are now possible to handle and distribute quickly. This is paramount in providing a global market, getting freshness and quality from harvest to consumer. Processing technologies are also more flexible and convenient nowadays. Processes like IQF and MAP permit much more flexible and convenient packaging and storage of seafood products. IQF allows consumers to use what they need without needing to defrost the whole package, which reduces waste and increases convenience. MAP extends the shelf life of seafood, thus removing the need for preservatives and allowing for more natural and minimally processed products.

Another important benefit associated with the contemporary ways of processing seafood is the aspect of environmental sustainability. Cryogenic freezing and other methods, like MAP, can aid in eliminating chemical preservatives and additives from seafood, minimizing the negative impact of seafood processing on the environment. Besides, modern packaging materials and technologies have been designed to be more environmentally sustainable through the use of recyclable and biodegradable materials, reducing waste, and preventing environmental pollution.

5.3. FREEZING AND THAWING OF SEAFOOD

Freezing and thawing are critical steps in the seafood industry concerning the preservation of quality, safety, and nutritional value from harvest through consumption. These processes affect seafood texture and flavor, as well as the overall quality. Best practices should be adhered to attain optimal results.

Freezing reduces the temperature of seafood to below 0°C, thus stopping microbial growth and enzymatic activity responsible for spoilage. Of the main advantages of this process, perhaps the most important one is that freezing retards the loss of freshness in seafood to a very great extent, thereby making it possible to store and distribute it over longer periods. The method and speed of freezing, however, have always turned out to be critical factors in the quality of the final product. Rapid freezing techniques, such as blast freezing and cryogenic freezing, are preferred because they help prevent large crystal growth that might inflict damage on the cellular structure of seafood. Large ice crystals that form in the slow freezing process normally puncture cell walls, resulting in a loss of moisture called drip loss, texture degradation, and a decrease in sensory quality upon thawing.

For seafood products that are sensitive in terms of texture and flavor i.e., fish fillets and shrimps, cryogenic freezing tends to work well because of the very low temperatures involved and high freezing rates. This method makes sure that ice crystals formed are small, thus minimizing cellular damage. Another very potent technique is blast freezing, where seafood is exposed to very strong cold air currents that rapidly lower the temperature, thus sustaining freshness and quality.

Thawing, or returning the frozen seafood to the unfrozen state, is equally important and should be carefully managed to prevent loss in quality. Uniformity of thawing and a controlled rate of thawing while avoiding temperature changes that will result in partial thawing and refreezing are some of the prime considerations in thawing. If thawing is done too rapidly, the outer layers thaw and deteriorate while the inside remains frozen, thus leading to an uneven texture and quality.

One of the good practices for defrosting seafood is to thaw it slowly under controlled refrigeration conditions, usually just above freezing temperatures at 0 – 4°C. This avoids uneven thawing of the seafood and maintains its juices and texture. A further method is cold water thawing in sealed packaging that does not allow water infiltration or contamination. Another alternative for small portions is microwave defrosting, but that should be carefully done to avoid partial cooking.

The cold chain has to be maintained from the freezing process right through to thawing. The term 'cold chain' refers to the series of temperature-controlled environments that keep the frozen seafood frozen from the time it is frozen until it reaches the consumer. If the cold chain is broken, the temperatures fluctuate, and so the seafood will start to thaw and then refreeze with disastrous results on its quality. Another good practice is the reduction, as much as possible, of exposure of seafood to temperatures at which microbial growth occurs. This means that there should be a fast transfer of seafood from freezing to storage and then from storage to the point of thawing to reduce the risk of microbial contamination. Furthermore, seafood should be handled with maximum care both at freezing and thawing to ensure avoidance of any physical damage, causing texture alterations or changes in appearance.

The freezing and thawing processes also have much to do with the packaging. Proper packaging materials protect seafood from freezer burn by keeping out air that dehydrates and oxidizes seafood. Vacuum packaging voids the air in the package, making the action of freezer burn less likely to affect the quality of the seafood.

5.4. SMOKING AND CURING OF SEAFOOD

Smoking and curing are two of the oldest methods of seafood processing, dating back several 100 years, not just as a means of preserving fish but to produce flavor. Today, these techniques remain in vogue because of their preservative effects and the taste and texture they give. The smoking and curing processes vary tremendously and produce different flavor profiles and major health considerations that must be understood both by the producer and the consumer (Figure 5.5).

Figure 5.5. *Smoked fish.*

Source: By Anne Burgess, CC BY-SA 2.0, https://commons.wikimedia.org/w/index.php?curid=13004451.

It involves the exposure of seafood to the smoke coming from burning wood that imparts flavor and acts as a preservative. Smoking reduces the moisture content in food while introducing antimicrobial compounds (Arvanitoyannis and Kotsanopoulos, 2011). There are two major types of smoking processes known as hot smoking and cold smoking. Hot smoking involves the cooking of seafood by the same process at temperatures that most commonly fall between 65°C and 85°C, which is equivalent to 150°F and 185°F, respectively. This method gives a firmer texture and a fully cooked product that can be ready to eat without further reconstitution. Smoked salmon, trout, and mackerel are some of the common examples of hot-smoked seafood.

In cold smoking, the seafood is smoked at low temperatures, normally below 30°C. It does not cook the fish but gives a delicate flavor of smoke to it while the raw texture of the seafood remains. Other cold-smoked products, such as lox or gravlax, could add curing processes to be considered safe and well-preserved. Most cold-smoked seafood is used as raw materials and is not recooked; therefore, the quality of fresh fish and strict hygiene standards should be highly regarded in processing.

Curing is often combined with smoking and involves the application of salt, sugar, or their mixture to seafood. Drying is the removal of moisture, either through surface drying by hot air or through the use of salt. In the curing process, osmosis draws out moisture from the fish, thus depriving growth for the microorganisms. Some methods of curing include dry and wet methods. The dry method directly rubs onto the fish a mixture of salt and sugar. Another is wet curing, where seafood is immersed in brine solution. Thus, curing prolongs the shelf life of seafood and gives flavor and texture to the final product. Hence, it is one step in processing things like gravlax, biltong, and salted cod.

The flavor profiles achieved through smoking and curing are highly dependent on several factors, including the wood used for smoking, the duration of time both for smoking and curing, and the exact recipes for brines or dry rubs. Most types of wood used in smoking originate from trees, particularly oak, hickory, maple, and apple, which lend distinctly different flavor profiles, ranging from robust and earthy to sweet and mild. With smoking and curing methods, this could result in a large variety of flavors based on regional and cultural tastes.

While smoking and curing are methods that render seafood both tasty and well-preserved, there are also some health considerations associated with them. One concern is the formation of potentially harmful compounds, like polycyclic aromatic hydrocarbons (PAHs) and nitrosamines, when smoking is done. These compounds can be formed through interaction between organic material from wood smoke and protein and fat in fish at high temperatures. Control of the conditions for smoking, primarily in terms of temperature and duration, and using clean, untreated wood can help minimize these risks. These compounds likely have maximum limits set for smoked foods by regulatory guidance; such limits ensure the safety of the foods.

Another health factor to consider is the amount of sodium in cured seafood. The salt content used in these processes is rather high; thus, it creates a potential health hazard for some, especially those suffering from hypertension or disorders of the cardiovascular system. In fact, manufacturers are increasingly working on low-sodium curing techniques and using alternative preservatives in these processes to still maintain quality and safety regarding such seafood.

5.5. FERMENTATION OF SEAFOOD

Fermentation is one of the oldest processing methods for seafood, many centuries old in several cultures. Fermentation not only preserves but also enhances flavors and improves the nutritional profile of fish and crustaceans. Fermented fish and crustaceans are food products made through the action of microorganisms, mostly lactic acid bacteria, which have converted the sugars and other substrates into acids, mostly lactic acid, alcohol, and other compounds. The process not only allows for the extension of the shelf life of seafood but unique flavors and textures are imparted, making fermented seafood in itself a delicacy in cuisine in large parts of the world.

5.5.1. Types of Fermented Seafood Products

Different cultures have huge diversities in products of fermented fish, with each culture having its own methods and recipes. One of the most famous fermented fish products is fish sauce, the usual seasoning in Southeast Asian cuisine, made by fermenting fish in salt for several months or even years. Through fermentation, fish proteins get broken down into amino acids, making fish sauce extremely savory, a characteristic flavor component in many dishes. The liquid that drains from this process is bottled and sold as fish sauce.

Another popular fermented seafood product is surströmming, a traditional Swedish dish made from fermented herring (Nygaard, 2019). The herring are lightly salted only to prevent complete spoilage before they are stored in barrels to ferment for several months. In the fermentation process, strong-smelling compounds are produced, highly appreciated by aficionados despite or maybe because of their pungent smell. In Korea, there is a category for fermented seafood known as jeotgal that includes different kinds of fish, shellfish, and roe. Jeotgal is used for seasoning or condiment and is often added to other fermented dishes such as kimchi. Some common kinds of jeotgal include myeolchijeot, which are fermented anchovies; saeujeot, or fermented small shrimps; and ojingeojeot, which are fermented squid.

5.5.2. Nutritional and Probiotic Benefits

Through the process of fermentation, complex molecules in seafood break down into simpler compounds, making digestion and absorption easier. For example, proteins in fish break into peptides and amino acids, thereby increasing their bioavailability. This, therefore, makes fermented seafood a great supply of quality protein with easy utilization by the body.

Moreover, during the fermentation process, a lot of bioactive compounds are formed, which are vitamins, antioxidants, and peptides with antimicrobial properties. Most fermented fishery products are good sources of B vitamins, especially B12, important for the normal functioning of nerve cells and blood cells. The antioxidants that are present within these foods protect the body from oxidative stress and inflammation and thus offer protection for health and well-being.

Among the most significant advantages of fermented seafood is its probiotic content. Probiotics are live microorganisms that provide health benefits to the host by the adequate amount of intake. Lactic acid bacteria, often isolated from fermented foods, have long been associated with the probiotic activity. Such beneficial bacteria can influence gut health, which in turn can be achieved through several means, such as balancing the intestinal microbiota, enhancing digestion, and boosting the immune system. Consuming fermented seafood regularly can help maintain a healthy gut flora, which is linked to various features of good health, including improved digestion, enhanced immune response, and even mental health benefits.

For individuals who are lactose intolerant or avoid dairy foods, products like jeotgal and fish sauce can supply an alternative source of probiotics. During fermentation, anti-nutritional factors levels drop; examples include phytic acid, which chelates with metals like iron and zinc to prevent their absorption into the body system. The bioavailability of nutrients increases through the reduction in such factors, hence the enhancement of the nutritional impact of fermented seafood. Apart from the nutritional value they offer, fermented seafood products also provide food security by minimizing losses. This means that fish and seafood are fermented, which otherwise would have been lost or gone unsalted. This aspect of sustainability assumes a higher significance in a world that is plagued by a lot of food waste.

5.6. DRYING AND DEHYDRATION OF SEAFOOD

Drying and dehydration are some of the oldest but still widely applied methods for seafood preservation. Basically, it is a process where moisture is removed from fish and other marine products to prevent micro-organisms and enzymatic activities from causing spoilage (Nguyen, Arason, and Eikevik, 2013). In doing this, drying and dehydration reduce the water activity and, at the same time, increase the shelf life and maintain quality. Drying and dehydrating methods used on seafood vary, and these different uses have different effects on the quality and shelf-life of the products.

5.6.1. Methods and Their Applications

Probably the oldest and most primitive technique of seafood drying is solar drying. It is the spreading of cleaned and salted fish or shellfish under the sun until the required level of moisture is achieved. Sun drying is low in cost and easy to perform, particularly in countries with abundant sunshine. It is mainly used for small fish, shrimp, and some categories of shellfish. This system, however, is very weather-dependent and occupies a lot of space and time, which consequently limits its efficiency and regularity.

Air drying, sometimes referred to as wind drying, is another method where seafood is hung in a well-ventilated place and dried due to natural air circulation. It is practiced in colder regions where sunlight is not that intense. Stockfish and klippfisk are dried cod products that are popular in Scandinavian countries. Air drying helps maintain the fish texture and flavor in this food, which has long been a significant part of local traditional dishes.

More extensive control is required in oven drying and smoke drying, which use an artificial heat source. Oven drying involves setting seafood in an oven at low temperatures to dry gradually. This method provides better control over the drying conditions and can therefore produce a more uniform quality. Smoke drying combines the drying process with smoking, where seafood is exposed to both heat and smoke from burning wood. This process dehydrates fish and gives it a smoky flavor simultaneously, hence improving the taste of the product and enhancing its shelf life.

Freeze drying (lyophilization) is a new method of dehydrating seafood where it is first frozen and then the pressure surrounding it is lowered so that the frozen water in it sublimes, going directly from the solid to the gas phase. Freeze-dried seafood has much of its original flavor, texture, and nutritional value because the low temperatures involved give little change in the structure of the food. This method is widely used for high-value seafood products like shrimps, scallops, and premium fish fillets. Freeze-dried seafood has an extremely high shelf life and low weight, thus being perfectly suitable for space-limited storage and long-distance transportation.

Spray drying and drum drying represent industrial dehydration methods used in the manufacturing of powder form seafood products like fish protein powders and seasonings, based on seafood. Spray drying is when the liquid seafood mixture is atomized or sprayed into a heated air chamber where the moisture is evaporated rapidly, creating fine powder particles. In drum drying, seafood purée is spread over the surface of a heated drum that will rapidly dry it as it rotates. Both methods provide highly concentrated and shelf-stable products that are easily stored and used in a wide array of food applications.

5.6.2. Impact on Shelf Life and Quality

Seafood, when dried appropriately, can be stored for many months or years, depending on the method applied and the conditions under which it is stored. In this way, the dried seafood becomes very useful as a source of food during seasons when fresh fish are not available and in areas where accessibility to fresh fish is low.

However, the drying process may have an impact on seafood quality in a lot of ways. The most noticeable change is in the texture. Dried seafood is much firmer and occasionally even chewy. The reason it is either desirable or undesirable depends on the culinary use. For instance, the firm texture of dried fish, upon its rehydration in soups and stews, adds variety to it; at the same time, the chewiness of squid makes it very popular as snacks.

Flavor changes also occur during drying. The concentration of flavors can enhance the taste of dried seafood to be more intense and savory. More specifically, smoke drying adds a peculiar smoky flavor, which is highly prized in many cuisines. Off-flavors are formed, though, or the delicate flavors found only in fresh seafood are lost if drying or improper drying conditions are excessive.

Another important point to consider is the nutritional value. While in general, drying preserves most of the protein and minerals found in the seafood, some of the vitamins may break down—especially the heat- and light-sensitive ones. This makes freeze-drying the best method concerning nutritional value retention in seafood since it uses low temperatures and gentle processing conditions. More pronounced loss of nutrients could occur via methods like sun drying or oven drying if the drying process is not well controlled.

5.7. IRRADIATION OF SEAFOOD

Irradiation is one of the modern processes in the seafood industry for improving safety and extending shelf life. It is a process of treating seafood with ionizing radiation i.e., gamma rays, X-rays, or electron beams, to kill or reduce the microbial pathogens, parasites, and spoilage organisms in seafood. One of the main objectives of irradiation is to ensure the safety of seafood by reducing the risk of foodborne illness, and it should be done without affecting product quality.

The principle of the irradiation process is based on the disruption of DNA and cellular structures of microorganisms, breaking them down to such an extent that they either die or become incapable of replication. In this way, it will be highly efficient against a variety of pathogens: Salmonella, *Listeria monocytogenes*, Vibrio species, and parasites such as Anisakis. This does not raise the temperature of seafood significantly, as opposed to what thermal treatments do, thereby preserving most of the sensory qualities, including texture, flavor, and nutritional value.

Major concerns about applying irradiation include these applications and uniformity to achieve the required microbial reduction without influencing quality. Irradiation dosage must be controlled in such a way that an overdose of radiation will cause adverse changes in texture and flavor. The seafood should be well-packaged before irradiation to avoid re-contamination of the processed products. Only those packaging materials shall be used that are compatible with irradiation and do not, by the action of ionizing radiation, release any substances that might constitute a health hazard.

The regulatory aspects related to irradiated seafood are managed through various agencies both at the national and international levels. In the United States, the Food and Drug Administration regulates the use of irradiation for seafood, specifying conditions including permitted doses and labeling requirements. Similarly, the European Food Safety Authority regulates the process of product irradiation within the European Union and thus verifies that seafood irradiated by this process is following food safety and quality. At the international level, the Codex Alimentarius Commission, established by FAO, the Food and Agriculture Organization, and WHO, the World Health Organization, provides guidelines and standards for the irradiation of foodstuffs and seafood. Thus, the regulatory structures in place provide a guarantee of the safety, efficiency, and clarity of this process to consumers.

Consumer acceptance of irradiated seafood varies greatly and is determined by many factors that include awareness, education, and cultural perceptions. To a large extent, the biggest concern among consumers has to do with the misconception that irradiated food is radioactive. This idea should be propagated: irradiation does not make the seafood radioactive because the ionizing radiation used does not remain in the food. It is through education campaigns and transparent labeling that these types of misconceptions are currently being averted across the board to assure consumers of the benefits and safety of irradiated seafood. Labeling is an important facet of consumer acceptance. Regulations in most countries stipulate that products of irradiated food be labeled by the international radura symbol and include a statement that the product was treated with ionizing radiation. This clarity will let consumers decide which seafood to buy. It could also focus more on benefits to consumers regarding irradiation, like improved safety, reduced spoilage, and extended shelf life, to gain consumer acceptance.

Although the advantages brought about by irradiation are obvious, this process is far from being accepted by the seafood industry. Large enough are the costs of irradiation facilities and compliance with various regulations that prevent some producers from adopting the technology. Nevertheless, in the long term, benefits from enhanced food safety, reduced wastage, and increased access to markets will compensate for the initial investments. Moreover, the more positive consumer perception and demand for irradiated products, the more producers will show interest in this technology.

5.8. PACKAGING TECHNOLOGIES FOR SEAFOOD

Packaging pertains to the retention of quality, safety, and shelf life of seafood. In this respect, development concerning new packaging materials and techniques has to a great extent increased the preservative ability of seafood products against spoilage, contamination, and physical damage. The developments ensure seafood remains fresh and appealing to consumers from the point of production to the moment of consumption.

Probably the most interesting innovation concerning packaging materials is the development of modified atmosphere packaging. Modified atmosphere packaging is based on changing the gas composition inside the packaging to slow down microbial growth and oxidation processes. Replacing oxygen with gases such as nitrogen and carbon dioxide, MAP substantially extends the shelf life of seafood. This technology works well with fresh and packaged products, such as fish fillets or shellfish, where authentic texture and flavor need to be sustained. Moreover, MAP can maintain the sensory characteristics of seafood while reducing the use of added chemical preservatives. It gives the customer a chance to sell their products to health-oriented consumers.

The other main development is vacuum packaging, where the air from a package is removed and then sealed tightly. This technique creates an anaerobic environment where the growth of aerobic spoilage microorganisms or oxidation cannot occur. Vacuum packaging is applied to a large number of fresh and frozen seafood products. It becomes very useful for products such as smoked salmon and ready-to-eat seafood, where quality and no spoilage are imperative. The tight seal will also protect the seafood from physical damage and dehydration, so it will be delivered to the consumer's hands in optimum condition.

Active packaging is an upcoming technology that involves incorporating certain substances in the packaging material, which will interact with the seafood to increase its shelf life. These may include oxygen scavengers, moisture absorbers, and antimicrobial agents. Oxygen scavengers remove the residual oxygen in a package, reducing the oxidation process and thus preventing spoilage. Moisture absorbers regulate the quantity of humidity in a package and prevent the growth of mold and other sorts of moisture-related spoilage. Antimicrobial agents directly inhibit the growth of pathogenic and spoilage microorganisms within the package and enhance the safety and shelf life of the product. Active packaging is an active solution to seafood preservation, as the kind of handling it provides changes with the product demands and storage/transport conditions.

Another advanced method in intelligent packaging involves integration with sensors and indicators, which monitor the condition of the seafood in the packaging material. The sensors monitor changes in temperature, humidity, and gas composition and could, therefore, offer real-time updates on freshness and safety. For example, TTIs could show if the seafood had undergone a temperature variation from what was recommended, something that would result in lower quality. Freshness indicators may indicate spoilage compounds, such as amines and sulfides, to signal probable quality problems to the consumer or retailer. On the other hand, intelligent packaging improves traceability and transparency to ensure the seafood stays safe and high in quality along the entire chain.

Sustainable packaging in the seafood industry is increasingly critical as buyers and producers seek to reduce their 'ecological footprint.' Biodegradable and compostable materials derived from renewable resources e.g., plant fibers and bioplastics, make for a very good alternative to traditional plastics. These can break down naturally after disposal and thus reduce landfill burden, minimizing pollution. Moreover, sustainable packaging aligns itself with the rising consumer demand for environmentally responsible products, therefore improving the brand image and market appeal of sustainable seafood companies.

5.9. SAFETY CONSIDERATIONS IN SEAFOOD PROCESSING

The safety of seafood products is of high relevance for the seafood industry, associated with a potential risk of contamination and various strict regulatory requirements in force. Contamination of seafood processing may emanate from microbial pathogens, residues of applied chemicals, and environmental pollutants. Thus, effective prevention and control measures are needed to minimize such risks and assure consumer health protection.

Of all processing aspects, microbial contamination poses one of the greatest safety challenges to seafood processing. Some of the major aquatic environment pathogens or those introduced during handling and processing include Salmonella, *Listeria monocytogenes*, Vibrio species, and Norovirus. Therefore, good hygiene practices should be put in place along the supply chain. It can also involve proper sanitation of facilities and equipment, regular testing for the presence of microbial contaminants, and following good manufacturing practices. As a requirement, most countries, including the United States and the European Union, necessitate Hazard Analysis and Critical Control Points systems. Basically, HACCP entails the identification of critical points in the processing chain where hazards may arise and then putting controls in place to prevent contamination.

Chemical contaminants such as heavy metals, pesticides, and antibiotics are major health hazards. Such substances accumulate in seafood through environmental exposure or even from their misuse in aquaculture. Monitoring programs are critical to ensure seafood products comply with the regulatory limits on chemical residues. This includes periodic examination of raw materials and finished products and sourcing of seafood from suppliers who have adopted sustainable and responsible practices.

Two of the very prominent emerging seafood safety concerns are microplastics and persistent organic pollutants. These contaminants enter the food chain and cause long-term health risks; control of this would require a mix of regulation-based enforcement and environmental protection measures, coupled with consumer education to reduce pollution and assure the long-term sustainability of seafood resources.

Assuring seafood safety is the most important role of regulation compliance. Needs regarding seafood safety are available from agencies like the Food and Drug Administration in the USA, the EFSA in Europe, and the Codex Alimentarius Commission. To meet these regulations, a system that includes rigorous testing, documentation, and traceability should be adhered to track every seafood from harvest to sale. This can ensure the consideration of all safety standards about seafood and easily detect anomalies so that correction measures are done.

Chapter 6

Nutritional Value and Health Benefits of Seafood

Contents

INTRODUCTION

For generations, seafood has been hailed as one of the basic ingredients of a healthy diet. It is more nutritionally dense and has manifold health benefits. This chapter aims to discuss seafood's nutritional value and its beneficial impact on human health. Ranging from the icy seas of the Arctic to the tropical seas, the great mass of seafood variety makes up a set of nutrients fundamental in everyday life for the maintenance of optimal health and prevention of chronic diseases.

Probably one of the most compelling arguments for consuming seafood is its high omega-3 fatty acid content. Those long-chain omega-3s, particularly eicosapentaenoic acid (EPA) and docosahexaenoic acid (DHA) contained in seafood, provide two very critical nutrients to help in cardiovascular and cognitive health. In contrast to the plant-based sources of omega-3s, the forms found in seafood are highly bioavailable and play an important role in reducing inflammation and supporting brain function.

Besides omega-3 fatty acids, seafood is an excellent source of high-quality protein. It delivers all the essential amino acids in proportions required by the human body for muscle repairs, immune function, and therefore overall growth (Lund, 2013). Seafood protein is very easily digestible; hence, it is of particular benefit to people right from infants to the elderly.

Seafood also contains essential vitamins and minerals, generally performing several important functions physiologically. The food group is extremely rich in vitamins like B12, D, and A and minerals such as iodine, selenium, and zinc. These have numerous benefits, for instance, bone health thyroxin synthesis, and immune support.

Furthermore, antioxidants in seafood make sure to provide a fight against oxidative stress and inflammation, which is linked with preventing chronic diseases such as heart disease, cancer, and diabetes. Natural antioxidants within seafood give it even greater value in nutritional power and a more potent promoter for long-term health.

Another important attribute of seafood is its low caloric content, making it an important component of a diet for individuals who intend to maintain weight without compromising on intake of nutrients. Since it contains high protein and low calories, seafood promotes a healthy metabolism and provides satiety, both important in weight management and obesity prevention. Seafood has been claimed as one of the food groups associated with reduced risks of heart attacks, strokes, and age-related cognitive decline. Moreover, regular consumption has been said to bring about improved cardiovascular health, enhanced cognitive function, and better developmental outcomes in children. Besides, the food group is a good source of nutrients that support fetal development during pregnancy and are also responsible for better health outcomes in infants and young children.

Seafood is more and more regarded as a functional food, delivering health benefits in addition to simple nutrition. Functional foods have bioactive compounds in them that will improve health, and seafood fits perfectly into the category of food containing a host of nutrients and compounds that benefit health. It's also important to be aware of the risks of seafood consumption, such as mercury and other heavy metal exposure, and the risk of allergic reactions. Therefore, understanding these risks and how they can be mitigated by proper choices is quite important in setting dietary guidelines for maximizing health benefits from seafood while minimizing harm.

6.1. OMEGA-3 FATTY ACIDS IN SEAFOOD

Omega-3 fatty acids belong to the family of essential polyunsaturated fats associated with health benefits. It is called "essential" because the human body cannot manufacture or synthesize it; hence, it has to be obtained from dietary sources. Seafood is one of the richest sources of these omega-3 fatty acids. Eicosapentaenoic acid, commonly called EPA, and docosahexaenoic acid (DHA) have the highest bioavailability from seafood and represent a wide range of health benefits (Figure 6.1).

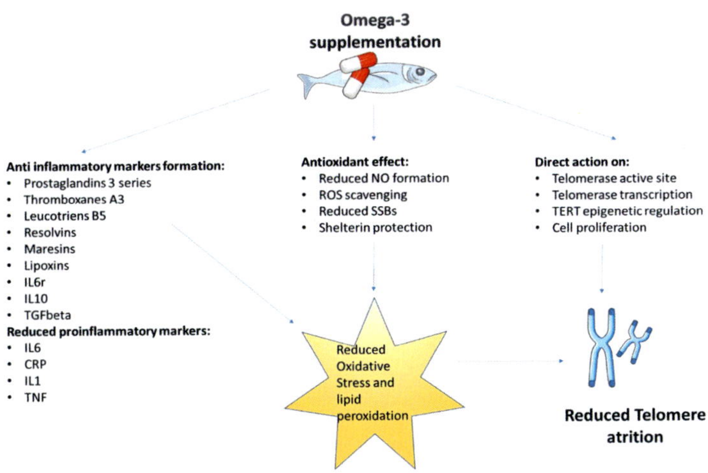

Figure 6.1. *Fish omega-3 fatty acids.*

Source: Ogłuszka M, Lipiński P, Starzyński RR. Effect of Omega-3 Fatty Acids on Telomeres—Are They the Elixir of Youth? Nutrients. *2022; 14(18):3723. https://doi.org/10.3390/nu14183723.*

6.1.1. Types and Sources

The seafood omega-3 fatty acids are generally long-chain, with high amounts of fatty acids in fatty fish and a variety of shellfish. Typical sources are salmon, mackerel, sardines, herring, anchovies, and tuna. These fish obtain omega-3s by eating phytoplankton and other small fish that are richly endowed with these fatty acids.

- Salmon, especially wild-caught ones like Alaskan and sockeye salmon, represent very good sources of EPA and DHA. A serving of 100 grams of salmon can provide upwards of 2 grams of omega-3 fatty acids, surpassing most individuals' daily recommended intake (Jensen et al., 2012).
- Another great source is mackerel, which contains about 1.5 grams of EPA and DHA in every 100-gram serving. What's more, the fish not only holds a ton of omega-3s but is also full of other very useful elements like vitamin D and selenium.
- Other powerful sources include sardines and herring, mostly taken canned or smoked. These are small fish, thus with lower levels of mercury, making them quite safe to take regularly. Anchovies are also major dealers in omega-3, often used for cooking due to their strong flavor.
- Tuna is a good source of omega-3s, although it should be consumed in moderation since the larger tuna species contain more mercury. Shellfish, such as oysters, mussels, and crab, also make some contribution to intake, although this is much smaller compared with oily fish.

6.1.2. Health Benefits

Omega-3 fatty acids have a huge list of health benefits, from physical to mental health. The most important ones are those associated with cardiovascular health. EPA and DHA lower triglycerides, blood pressure, and the risk of arrhythmias (Schacky and Harris, 2007). They contribute to a reduction in inflammation, one of the grounds for atherosclerosis, which means reducing the risk of heart attack and stroke.

Omega-3 fatty acids are also indispensable for brain health. Of these, DHA is considered one of the major structural components of the brain and retina. Adequate intake of DHA has therefore been related to improved cognitive function and a reduced risk of neurodegenerative diseases like Alzheimer's. For infants and children, it is essential in brain development and cognitive performance. Many experts advise pregnant and lactating women to take sufficient amounts of DHA for the proper brain development of the fetus and infant.

The other significant area where omega-3 fatty acids come into play relates to their mental health benefits: they alleviate symptoms of depression and anxiety. Mechanistically, EPA and DHA are thought to influence the activity of neurotransmitters and neuroinflammation, key modulators of mental health disorders.

Besides, omega-3 fatty acids help to care for your eyes. They prevent age-related macular degeneration and dry eye syndrome by reducing inflammation in ocular tissues. Anti-inflammatory activity positively affects the retina and all other ocular tissues.

Omega-3s will also help you with joint health. The anti-inflammatory effects of them contribute to alleviating the symptoms of rheumatoid arthritis and other inflammatory joint disorders. Intake of omega-3s regularly can help improve flexibility in the joints and reduce their pains.

6.1.3. Dietary Recommendations

Due to the number of health benefits from omega-3 fatty acids, various health organizations have specific dietary recommendations for adequate intake. According to the American Heart Association, at least two servings of fatty fish per week are recommended, equivalent to about 250–500 milligrams of EPA and DHA per day. This intake amount has been associated with a large decrease in CVD risk.

In individuals with pre-existing blood flow issues in the heart, higher doses are prescribed—one gram per day of EPA and DHA. Supplements can be taken by people who do not eat enough fish, but usually, it is preferred to use whole food sources that have added nutrition.

Pregnant and breastfeeding women should consume at least 200 milligrams of DHA per day to support fetal and infant brain development. This may be achieved with seafood consumption and intake of foods or supplements enriched with DHA.

6.2. PROTEIN QUALITY OF SEAFOOD

Seafood is one of the globally recognized sources of good protein and provides a lot of health benefits. Seafood contains not only abundant protein but also high biological value. It means that in terms of human utilization, it is highly efficient. It has a rather high biological value due to the presence of essential amino acids in proportions closely matching the human dietary requirement.

6.2.1. Biological Value and Essential Amino Acids

The biological value of a protein source is the efficiency with which the human body can use the ingested protein for growth and maintenance. In this respect, seafood scores highly since it contains a complete set of essential amino acids. Such amino acids cannot be synthesized by the body and thus need to be ingested in the diet. These include leucine, isoleucine, valine, lysine, methionine, phenylalanine, threonine, tryptophan, and histidine.

Proteins from fish and crustaceans are especially rich in essential amino acids. For example, salmon, tuna, and cod have a high content of lysine and methionine—two amino acids that make up a wide spectrum of physiological activities, from repair to metabolism in the body. High amounts of these essential amino acids are also present in shellfish like shrimp and crab, adding value to their overall nutritional profile.

Another advantage associated with seafood protein is its easy digestibility. The digestibility of protein refers to the degree to which the body can digest and subsequently absorb the amino acids involved. Comparatively, there is fairly good digestibility associated with seafood proteins in comparison to some other animal proteins. Thus, the body can make better use of the nutrients derived from these food sources. This is beneficial for those experiencing digestive problems or the elderly, whose digestion system may have declined.

But how do proteins from seafood compare with other protein sources?

- **Meat vs. Seafood:** While both are good sources of complete proteins that supply all essential amino acids, in general, seafoods have a higher digestibility score and can offer several other health benefits since they have less saturated fat and more omega-3 fatty acids. For instance, while beef and chicken are excellent sources of protein, they lack the cardiovascular health benefits seen in fatty fish such as salmon and mackerel due to their richness in heart-healthy omega-3s. Other seafood, like shellfish, contains a fully unique nutritional profile of its own, dense in minerals like iodine and zinc found less frequently in land-based meats.
- **Dairy Versus Seafood:** The very good sources of protein come from milk, cheese, and yogurt. These foods provide a large number of EAAs to be taken in by the body, specifically leucine, which initiates muscular synthesis. However, seafood frequently provides a greater variety of nutrient intake, including omega-3s and more significant amounts of vitamins D and B12. Furthermore, it is a very good source of complete protein for those who are lactose intolerant or allergic to dairy products and the ones who just abhor milk and

its products. Seafood is a fantastic alternative source of complete proteins for those who are lactose intolerant or allergic to dairy.

- **Plant-Based Proteins Versus Seafood:** The most popular plant-based proteins, mainly due to the environmental advantages and individual health benefits linked with them, happen to be legumes, nuts, seeds, grains, and so on (Saravana, Ummat, O'Donnell, and Tiwari, 2022). Plant proteins tend not to be complete proteins, meaning they lack one or more amino acids that are essential. For example, legumes are generally low in methionine, while grains naturally have rather low lysine content. Even if the combination of different plant proteins can yield a complete amino acid profile, seafood alone can already supply all essential amino acids in a single source. Besides, plant-based proteins generally have lower digestibility scores compared with those of animal origin, such as seafood. This makes seafood an extremely useful source of protein for those who wish to easily obtain all the essential amino acids.

This returns us to the overall health impact of various protein sources beyond the amino acid profile. Seafood is associated with a plethora of health benefits, including improved cardiovascular health, reduced inflammation, and enhanced cognitive function, primarily from omega-3 fatty acids and other bioactive compounds. On the other hand, red meat has been linked to higher risks of certain diseases when taken in excessive amounts, such as heart disease and colorectal cancer. This makes seafood a preferable choice for people looking to optimize their protein intake for better health.

6.3. VITAMINS AND MINERALS IN SEAFOOD

Seafood is an excellent source of vitamins and minerals essential to health and well-being. The nutrient makeup in seafood includes many micronutrients whose levels are key players in numerous physiological processes.

Among these vitamins, the most common one found in surplus in seafood is vitamin D. While most other food sources do not contain much vitamin D, fatty fish provide a large enough amount of the nutrient. Salmon, mackerel, and sardines are fatty fish that are rich sources of vitamin D to ensure calcium absorption for healthy bones and teeth. Provided the levels of this vitamin are adequate in a person's system, disorders such as rickets and osteoporosis can be prevented. Besides calcium metabolism, vitamin D is also responsible for proper immune functioning and has been associated with a reduced risk for many chronic diseases like certain cancers and cardiovascular diseases. Another key nutrient that seafood is rich in is vitamin B12. Vitamin B12 plays a crucial role in the formation of red blood cells, in neurological functions, and in DNA synthesis. Deficiency in vitamin B12 is usually associated with anemia, fatigue, and neurological disorders. Clams, oysters, and many species of fish are significant sources of this vitamin, which shows their importance, especially for the elderly and those with plant-based diets who are at risk of having deficiencies.

In addition to vitamins D and B12, seafood is also an excellent source of vitamin A. As the principal vitamin maintaining vision, it plays a crucial role in immune function and skin health. Fish liver oils, particularly cod liver oil, are considered very rich sources of vitamin A. The stable consumption of fish will allow for a sufficient level of this vitamin and give good support to the healthy functioning of the visual apparatus while decreasing the risk of night blindness and other vision failures.

Seafood is also a good source of several essential minerals, particularly iodine, selenium, and zinc. Iodine is involved in thyroid function by being part of the thyroid hormones responsible for regulating metabolic processes in the body. Deficiency of iodine is associated with goiter and hypothyroidism. Seafood, especially seaweed, shellfish, and some fish, is a natural and effective way to ensure a supply of iodine.

Selenium is another essential trace mineral found in seafood, known for its role as an antioxidant (Ralston, 2009). It protects the cells from oxidative damage and supports the immune function. Although notoriously high in Brazil nuts, seafood like tuna, halibut, and sardines are among the significant contributors of selenium intake to ensure that adequate antioxidant defenses are provided for within the human body.

Zinc, found in oysters, crabs, and other seafood, is involved in immune function, wound healing, DNA synthesis, and cell division. Sufficient intake of zinc enables the body to maintain a healthy immune system and promotes normal growth and development, especially during childhood. Iron is another important mineral that is found in seafood, especially in clams, mussels, and oysters. It helps the body manufacture hemoglobin, which carries oxygen in the blood. Intake of heme iron, the easily absorbed form of iron from seafood, prevents iron deficiency anemia and keeps energy levels and cognitive function intact.

Magnesium, mainly found in fish, especially in mackerel and salmon, is involved in more than 300 biochemical reactions in the body: from the functioning of muscles and nerves to the regulation of blood sugar levels and the synthesis of proteins. Sufficient amounts of magnesium in the diet are necessary for maintaining good cardiovascular health and preventing a range of diseases, from hypertension to heart disease.

6.4. ANTIOXIDANTS IN SEAFOOD

Antioxidants can be described as compounds that help balance out free radicals known to have destructive abilities at the cellular level in humans and culminate in chronic illness. Antioxidants work to protect the body from such damage through oxidation. Seafood is one of the high sources of such antioxidants, providing many bioactive compounds that would help give credence to their potential in enhancing health.

One of the major antioxidants derived from fish is selenium, a trace mineral endowed with potent antioxidant activity. It is one of the constituents of selenoproteins that are essential for maintaining cellular protection against oxidative damage. Seafood, particularly fish like tuna, salmon, and sardines, and shellfish like oysters and shrimp, is rich in selenium. Intake of such amounts of selenium-rich seafood may improve the general antioxidant defenses of the human system, thereby alleviating the development of oxidative stress-related diseases, such as cancer and heart diseases.

Astaxanthin is another potent antioxidant that, like other astaxanthin carotenoid pigments, can be found in large amounts mainly in seafood, particularly crustaceans such as shrimps, crabs, and lobsters, and sometimes in fish such as salmon and trout. This gives seafood items their pink-to-red color. In fact, astaxanthin has been reported as displaying very high antioxidant activity—far greater than any other known carotenoids or vitamins E and C combined. Astaxanthin protects tissue against such oxidative stress, forming the basis of various protective actions of use in cardiovascular health and skin protection.

Vitamin E, as found in fish oils and oily fish such as salmon, mackerel, and sardines, is an excellent fat-soluble antioxidant that protects cell membranes from damage to the cell caused by free radicals. It works synergistically with other antioxidants, especially vitamin C, in enhancing the overall antioxidant capacity. Vitamin E prevents the oxidation of LDL cholesterol and reduces the incidence of atherosclerosis and heart diseases. In addition, it also works for immunity and healthy skin.

Another very important mineral with antioxidant activity is zinc, found in seafood: oysters, crab, and lobster. This is the element constituting various enzymes that protect from oxidative stress and participate in DNA repair and immune function. Enough intake of this mineral from zinc-rich foods ensures healthy skin, promotes wound healing, and probably protects against such chronic diseases whose origins are rooted in oxidative damage.

Polyphenols, also an antioxidant is available in seafood, mainly those that consume algae, such as some species of shellfish and fish. They have been shown to decrease inflammation, improve blood vessel function, and offer some protection against cardiovascular diseases. Polyphenols also work in support of brain health through a reduction in oxidative stress and inflammation in neural tissue.

Through antioxidants, free radicals can reduce the risk of chronic illnesses such as cardiovascular diseases, cancer, and neurodegenerative disorders. Astaxanthin and polyphenols also modulate anti-inflammatory actions, protecting against heart diseases by attenuating inflammatory processes within the blood vessels and improving lipid profiles.

Antioxidants in seafood also work to improve immune function. The role of selenium, vitamin E, and zinc in the immune system spans from the production of immune cells down to protection from oxidative damage, which helps raise the resistance of the body to infections and reduces the incidence of immune diseases.

They also have a property that enables them to assist seafood in promoting healthy skin. The compounds astaxanthin and vitamin E protect the skin against oxidative damage arising from UV and environmental pollutants. This ensures that it prevents premature aging, reduces the risk of cancers of the skin, and maintains the integrity of the skin in general.

6.5. LOW-CALORIE NATURE OF SEAFOOD

While seafood is considered very dense, in terms of its nutrient content, it is very low in calories, making it an ideal addition to diets aimed at weight loss and general nutrition.

6.5.1. Caloric Content and Weight Management

While the calorie count varies from one type of seafood to another, further dependent on the method of preparation, it is generally lower than most other animal protein products. A 100-gram serving of cooked fish will provide about 70–200 calories, again depending upon the species. Lean fish like cod, haddock, and flounder are very low in calories, typically providing fewer than 100 calories per serving. Even fatty fish, rich in omega-3 fatty acids, such as salmon and mackerel, are only moderately high in calories at about 200 per serving.

Other low-calorie yet rich in protein and essential nutrient sources include shellfish like shrimp, crab, and clams. A serving of 100 grams of shrimp contains approximately 99 calories, while the same amount of crab comes with around 87 calories. These low-calorie contents, along with the high protein contents, make seafood an ideal choice for those who want to keep their weight under control.

Low-calorie seafood has several benefits in terms of weight management. First, it is rich in protein, a more satiating nutrient compared to carbohydrates or fats. This will further keep a person full longer after eating seafood and thus reduce overall caloric intake and prevent overeating. With this in view, and the fact that protein has a higher TEF than fats and carbohydrates, it means the body uses more energy in the digestion and metabolism of protein, and hence this could help in weight reduction efforts.

Moreover, the high protein content in seafood will help in the maintenance of lean muscle mass while one is losing weight. While people reduce their caloric intake with the view of losing some weight, there is usually a risk of losing muscles together with fat. A diet that contains enough protein will always help the maintenance of muscles and ensure that weight loss comes from the fat stored in the body. This is essential for keeping the metabolic rate and general health during weight loss.

6.5.2. Benefits in Balanced Diets

Seafood can be easily incorporated into a balanced diet making it important for nutrition. For instance, fish is very versatile. It can be grilled, baked, steamed, or poached i.e., the different ways it can be cooked that can be healthy and not fatty. Unlike red meats, which are cooked in high-fat methods like frying or grilling with a lot of added fat, seafood can be made with very little added fat and still has flavor.

Aside from this, seafood will fit into many dietary patterns that place an overall emphasis on plant-based foods combined with healthy fats and lean protein sources, including the Mediterranean, pescatarian, and flexitarian diets. These diets show an association with several health benefits, such as risk reduction of chronic diseases, improved weight management, and better overall nutrient intake. Additional benefits associated with adding seafood to meals are related to weight management. When seafood is used in a balanced diet, this protein-rich food will supply all the important nutrients while keeping the intake of calorie count within bounds. A meal that incorporates grilled salmon, quinoa, and steamed vegetables, for example, yields a nutrient-rich, low-calorie meal for a person to be full up and ensures good metabolic health.

Finally, seafood intake is linked to stronger adherence to healthier patterns of eating. The richer the flavors and versatility of this group of foods make it much easier for people to enjoy balanced, healthy meals without feelings of restriction, which could lead to more sustainable dietary habits with better weight management outcomes in the long term.

6.6. HEALTH BENEFITS OF SEAFOOD CONSUMPTION

Seafood has acquired wide recognition for its huge health benefits. To some part, this is due to its nutrient-rich food value that represents a rich source of good protein, essential fatty acids, vitamins, and minerals. While we have individually discussed the benefits accrued to diverse nutritional components of seafood, it has enacted positive cardio-health and cognitive growth (Figure 6.2).

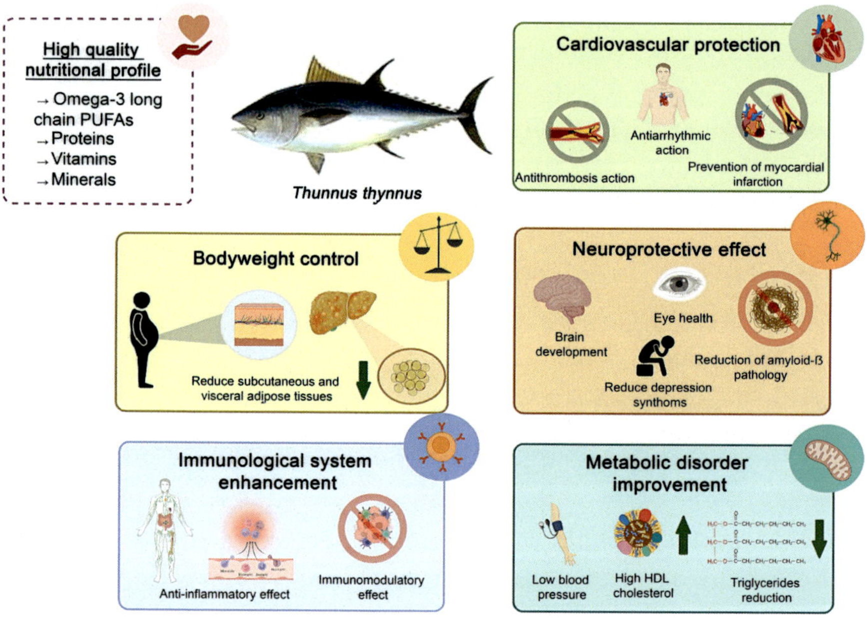

Figure 6.2. *Health benefits of fish consumption and its bioactive constituents.*

Source: Chamorro F, Cassani L, Garcia-Oliveira P, Barral-Martinez M, Jorge AOS, Pereira AG, Otero P, Fraga-Corral M, P. P. Oliveira MB and Prieto MA (2024); Health benefits of bluefin tuna consumption: (Thunnus thynnus) as a case study. Front. Nutr. 11:1340121. doi: 10.3389/fnut.2024.1340121.

6.6.1. Cardiovascular Health

Some of the most compelling reasons seafood should be part of the diet include its immense benefits for cardiovascular health. In fish, omega-3 fatty acids play a significant role in keeping hearts healthy. Unlike other sources of fats, omega-3 fatty acids are anti-inflammatory in nature, thus lessening the potential risks of heart diseases.

Omega-3 fatty acids contribute to heart health in many ways. Among them, they lower blood levels of triglycerides, one of the major risk factors for heart disease (MD et al., 2021). High triglycerides cause stiffening of arteries and increase the possibility of a heart attack or stroke. Consuming seafood rich in omega-3 regularly may considerably reduce the levels and promote a healthier blood lipid profile.

Besides, omega-3s have been shown to reduce blood pressure in subjects with hypertension. Among the main risk factors for heart disease and stroke, high blood pressure can be intervened through diet, which may emerge as a very potent tool. Anti-inflammatory effects of omega-3 fatty acids also reduce chronic inflammation, which does damage to blood vessels and often leads to atherosclerosis.

In addition, seafood consumption has been linked to a reduced risk of arrhythmias, generally defined as irregular heartbeats that could lead to sudden cardiac death. The omega-3 fatty acids in seafood help to stabilize electrical activity in the heart and thereby reduce the likelihood of developing arrhythmias. This protective effect is of further importance to people who already have heart disease or are at high risk of heart disease.

Apart from the omega-3 fatty acids, it also contains some other nutrients that are useful in maintaining heart health by providing vitamin D and selenium. Vitamin D helps lower blood pressure and promoting cardiovascular health, while selenium acts as an antioxidant that may reduce oxidative stress in the heart, lowering subsequent inflammation.

6.6.2. Cognitive and Developmental Benefits

There are cognitive and developmental benefits associated with seafood intake, more particularly about brain health and development. DHA is one of the most important components of omega-3 fatty acids found in seafood and is generally regarded as one of the principal determinants of brain development and function. It is essential for the formation of brain tissue; adequate intake is crucial for cognitive health throughout life.

Seafood is very essential for pregnant women due to the presence of EPA and DHA, which help in fetal brain and eye development. DHA contributes much to the neural development process; inadequate intake of it during pregnancy might result in less than optimal neurodevelopmental outcomes for the offspring. Intake of adequate amounts of DHA during pregnancy is generally associated with better performance of cognitive skills and visual acuity in children born to such mothers. In infants and young children, DHA continues to play a role in brain development. Breast milk is naturally a source of DHA; however, mothers who regularly eat seafood can better improve the levels provided to their breastfed infants. This can enhance cognitive development and support better learning and memory outcomes as the child grows.

It has been linked to many other health benefits in adults, including improved cognitive function and a reduced risk of neurodegenerative diseases such as Alzheimer's and dementia. Omega-3 fatty acids work to maintain the structure and functions of neurons, support synaptic plasticity, and reduce inflammation in the brain—effects that help to improve memory, attention, and overall cognitive performance.

In older people, an adequate intake of DHA and EPA from seafood consumption will help slow cognitive decline and sustain mental health. There is growing evidence that omega-3 fatty acids reduce symptoms of depression and anxiety by modulating neurotransmitter function and attenuating neuroinflammation. Seafood becomes a very valuable food in the support of mental health and maintenance of cognitive longevity.

Furthermore, neurotransmitters are manufactured from high-quality proteins present in fish and promote general nervous system health. Other nutrients like B vitamins, iodine, and zinc in seafood act to promote cognitive function and brain health. It is at this point that B vitamins play their role in energy metabolism going on in the brain; iodine functions in thyroid and brain development; and zinc has activity in neurotransmitter activity and cognitive stability.

6.7. SEAFOOD AS A FUNCTIONAL FOOD

Seafood is increasingly being recognized as a functional food, that is, foods that provide health benefits besides the nutritional value. Functional foods are foods that have bioactive compounds and can possess either health-promoting benefits or reduce the risk of chronic diseases. Being rich in nutrients with high-quality proteins, essential fatty acids, vitamins, minerals, and antioxidants, seafood can be considered an archetypal functional food.

Some examples of seafood with their functional components are:

- **Fatty Fish:** Salmon, mackerel, sardines—are extremely rich in omega-3 fatty acids, particularly EPA and DHA. Omega-3s play an extremely important role in heart and brain health, reducing inflammation, and improving cardiovascular function generally. Fatty fish also provide high-quality protein and essential vitamins like D and B12.

- **Shellfish:** Oysters, clams, and mussels are nutrient-dense, acting as a huge source of vitamins and minerals like zinc, selenium, and vitamin B12. These nutrients support immune function, antioxidant defenses, and neurological health. Besides, shellfish is an excellent source of taurine, an amino acid involved in heart health.
- **Seaweed and Algae:** These are normally consumed in Asian diets and are very rich in dietary fiber, iodine, and antioxidants such as fucoxanthin and phlorotannins. These compounds show anti-inflammatory and anti-cancer activities, exert thyroid function, and create gut health.
- **Crustaceans:** Like shrimp, crab, and lobster they have a rich content of astaxanthin, their red or pink carotenoid pigment. Astaxanthin is an efficient cell protector from oxidative stress, with the potential to improve skin condition and reduce inflammation.

6.7.1. Health-Promoting Properties as Functional Foods

- **Anti-Inflammatory Effects:** The risk of chronic inflammation is associated with a host of various diseases, including heart diseases, diabetes, and cancer. Omega-3 fatty acids are highly anti-inflammatory in nature and hence lower the levels of inflammation within the body. Other compounds present within seafood, such as astaxanthin and taurine, also exert anti-inflammatory effects that further add to the health benefits accruable from seafood.
- **Immune Support:** Vitamins and minerals in fish, particularly zinc, selenium, and vitamin D, play a critical role in immune support. Zinc is engaged in the processing of nearly all functions attributed to immune cells. Selenium acts as an antioxidant that protects the cells from damage, while vitamin D controls the response of the immune cells. Therefore, these nutrients aid in fighting infection and lowering the risk of chronic diseases.
- **Bone Health:** Vitamin D and calcium are two inseparable factors for the upkeep of bone health. Vitamin D is found in high levels in fatty fish, such as salmon and mackerel, which in general enhances calcium absorption and facilitates bone mineralization. Intake of seafood rich in vitamin D regularly will prevent osteoporosis and maintain strong bones.
- **Weight Management:** Seafood is generally a low-calorie, high-protein food—therefore, a perfect addition to any weight

management diet. High-protein foods cause satiety, creating the feeling of fullness, hence reducing overall calorie intake by keeping the individual fuller for longer. Moreover, the healthy fats, especially omega-3s, in seafood increase metabolism and help burn fat.

- **Skin Health:** Astaxanthin is an antioxidant of crustacean origin; omega-3 fatty acids are found in oily fish and are potent against oxidative damage to skin tissues by solar UV radiation and environmental pollutants. These compounds contribute to the elasticity of the skin, therefore improving the appearance of wrinkles and protecting the overall health of the skin.

6.8. RISKS ASSOCIATED WITH SEAFOOD CONSUMPTION

6.8.1. Mercury and Heavy Metals

One of the major problems linked to the consumption of seafood is mercury contamination. Mercury is a heavy metal that may prove neurotoxic; it especially affects fetal development and young children. Fish and shellfish absorb mercury from water in the most toxic form methylmercury and then it accumulates in their tissues. These are higher in larger and longer-living fish, such as shark, swordfish, king mackerel, and some types of tuna. The biomagnification of mercury can result in mercury poisoning from excessive consumption of fish, leading to neurologic and developmental problems (Storelli, 2008).

Health experts recommend reducing the intake of fish that contain a high level of mercury and increasing the intake of fish that contain a low level of mercury to minimize exposure. Some small fish, such as salmon, sardines, and trout, have relatively low levels of mercury and are regarded as safe, especially for expectant mothers, nursing mothers, and young children. Mixing up seafood can also help reduce the risk of excessive intake of mercury.

6.8.2. Persistent Organic Pollutants

Persistent organic pollutants are a group of environmental pollutants that include specific chemicals such as polychlorinated biphenyls, dioxins, and certain pesticides. These compounds have very high stability and may persist in the environment for a very long time. Since they bio-accumulate along the food chain, seafood, mainly fatty fish, may contain high amounts of POPs because of their fat solubility. Exposure to POPs has been linked to endocrine disruption, reproductive and developmental problems, immune system impairment, and increased incidence of certain cancers. Such risk of POP exposure can be reduced by choosing seafood from less polluted waters and by eating a variety of fish to avoid high intake levels of any single contaminant. The skin and the visible fat of fish can also be removed to reduce the intake of fat-soluble pollutants.

6.8.3. Allergenic Concerns

Seafood allergies are common and can cause severe reactions in susceptible individuals. Among the most prevalent food allergies are those to the shellfish, which include shrimp, crab, lobster, and other related species. Fish allergies are somewhat less common, although they can still be dangerous. The symptoms of seafood allergy can be mild, like hives or digestive discomfort, or serious like anaphylaxis, a potentially life-threatening condition that requires instant medical care.

People with diagnosed seafood allergies should avoid those seafood types that provoke their reactions. Cross-contamination in restaurants and food processing plants is also a potential problem, so individuals with serious allergies should inform food preparers about the existence and severity of their allergies when they eat out or buy seafood products. Required allergen labeling helps consumers recognize and bypass products that might contain an allergen to add another measure of protection.

6.8.4. Microbial Contamination

Consuming seafood, mainly raw or undercooked products like sushi, sashimi, and oysters, also comes with the risk of microbial contamination. There are a huge number of pathogens that are carried in seafood, such as Vibrio, Salmonella, and Listeria. The food poisoning caused by these agents in consumers leads to gastrointestinal disorders, fever, and even life-threatening conditions.

Attention has to be paid to handling and cooking seafood to reduce the risk of microbial contamination. The raw seafood should be procured from a credible supplier with strict safety measures in place. Generally, cooking seafood to an internal temperature of at least 145°F destroys most pathogens. In addition to proper storage and hygiene, such as refrigeration after time-to-time preparation and preventing cross-contamination with other foods, the risk of foodborne illness can be controlled.

6.8.5. Parasites

Seafood, in particular raw or half-cooked fish, may contain the parasites Anisakis and Diphyllobothrium. These parasites result in infections that present with gastrointestinal symptoms like nausea, vomiting, and abdominal pain. Freezing fish at particular temperatures for a specified period kills the parasites, lowering the risk of infection. For instance, the FDA recommends that fish be frozen at -4°F (-20°C) or below for seven days or to a temperature of -31°F (-35°C) or below until solid and then stored at a temperature of -31°F (-35°C) or below for 15 hours.

6.8.6. Environmental and Sustainability Issues

In addition to direct human health risks, there exist related environmental and sustainability concerns. Large-scale overfishing, bycatch, and habitat damage are some of the most important reasons that threaten marine ecosystems. It may thus also give rise to environmental problems through seafood consumption that originates from unsustainable sources and thence causes biodiversity impacts on the habitats of the ocean.

If consumers really want to help alleviate these concerns, they could opt for seafood that has been certified by programs like the Marine Stewardship Council, also known as MSC, and the Aquaculture Stewardship Council for sustainable catching or farming. These certifications indicate that seafood has been caught or farmed using methods and practices that least affect the environment and assure a long-term resource base.

6.9. DIETARY GUIDELINES FOR SEAFOOD CONSUMPTION

Dietary guidelines for seafood intake emphasize the health benefits associated with fish and seafood consumption but also highlight that such intake should be made in consideration of their potential contaminant risks. Specific recommendations have been advanced by health organizations, including the American Heart Association, the Food and Drug Administration, and the Dietary Guidelines for Americans, to advise on making informed choices.

The American Heart Association recommends that people eat fish, particularly the oily kinds like salmon, mackerel, and sardines, at least twice a week. That works out to about 8 ounces (two servings) per week of seafood. These types of fish are packed with omega-3 fatty acids, which have huge health benefits for good heart health. Omega-3s help reduce the chance of heart disease by lowering triglycerides, and blood pressure, and preventing arrhythmias. Pregnant and breastfeeding women should consume 8–12 ounces of seafood per week to support fetal brain and eye development, choosing low-mercury seafood like salmon, tilapia, and shrimp.

The Dietary Guidelines for Americans recommend at least 8 ounces of seafood per week in the diet for adults, based on a 2,000-calorie diet. This could be based on an individual's needs concerning caloric intake, age, and health condition. These recommendations really underscore the breadth of the recommendation and emphasize the need to include a variety of seafood types to have benefits with minimized risks from contaminants such as mercury and PCBs. Children should be encouraged to eat seafood while adjusting the serving size according to age and caloric intake.

Standard serving sizes are 3–4 ounces for adults, which is about the size of a deck of cards, while for children, serving sizes are smaller and the amount of food eaten depends on their needs. According to health organizations, serving sizes and their frequency should be adjusted according to personal dietary needs, taking into consideration factors such as age, weight, and general health.

It is better to eat a variety of seafood so that exposure to contaminants does not become too great, and to focus on choices with low mercury levels. Vulnerable groups, such as pregnant women, nursing mothers, and young children, are more susceptible to high levels of mercury found in fish like shark, swordfish, king mackerel, and some species of tuna. Safe choices are indicated through the fish consumption advisory and other resources offered by the FDA and EPA.

Chapter 7

Allergenicity of Seafood

Contents

INTRODUCTION

Seafood is an actively used component in most diets worldwide, providing a wide range of nutritional benefits. However, in some individuals, the intake of seafood could potentially kick-start severe allergic responses. Allergies to seafood are among the most widespread food allergies experienced by millions of people globally, manifesting in ways ranging from simply giving mild hives and gastrointestinal symptoms, to causing severe anaphylaxis that may endanger the patient's life. The proteins that lead to allergic reactions occur in these 2 major basins of fish protein source: fish and shellfish (Figure 7.1).

References	Location	Sample size	Age range (Year)	Methodology	Shellfish (rank)	Fish (rank)
Li et al. (8)	Hong Kong	6,194	6–11	sIgE > 0.7 kUA/L	4.7% (3rd)	1.2% (21st)
				SPT	2.43% (1st)	0.22% (4th)
				sIgE + symptoms	1.05% (1st)	0.2% (2nd)
	Guangzhou	5,542	6–11	sIgE > 0.7 kUA/L	2% (2nd)	0% (/)
				SPT	1.14% (1st)	0% (/)
				sIgE + symptoms	0.18% (1st)	0% (/)
	Shaoguan	5,139	6–11	sIgE > 0.7 kUA/L	13.1% (1st)	0% (/)
				SPT	4.53% (1st)	0.04% (10th)
				sIgE + symptoms	0.65% (1st)	0% (/)
	Bengaluru/Mysore	5,677	6–11	sIgE > 0.7 kUA/L	10.3% (1st)	0.4% (26th)
				sIgE + symptoms	0% (/)	0% (/)
Le et al. (9)	Vietnam	9,039	16–50	Self-reported	6.88% (1st)	3.71% (2nd)
				Doctor-diagnosed	2.95% (1st)	1.58% (2nd)
Dai et al. (10)	Wenzhou	4,151	3–6	Self-reported	2.46% (2nd)	1.35% (6th)
Ziyab et al. (11)	Kuwait	3,864	11–14	Questionnaire + Clinical history	1.3% (5th)	1.6% (3rd)
Le et al. (12)	Hue	4,443	2–6	Self-reported	5.22% (1st)	1.55% (2nd)
				Doctor-diagnosed	4.79% (1st)	1.37% (2nd)
	Tien Giang	4,177	2–6	Self-reported	4.29% (1st)	1.7% (4th)
				Doctor-diagnosed	2.8% (1st)	1.1% (4th)
Kim et al. (13)	Korea	29,842	6–16	Self-reported	0.84% (2nd)	0.32% (3rd)
Park et al. (14)	Seoul	16,749	0–6	Self-reported	0.5% (5th)	0.4% (6th)
Dey et al. (15)	Kolkata	5,161	All	Self-reported	3.43% (/)	4.59% (/)
Lao-araya et al. (16)	Chiang Mai	452	3–7	Self-reported	3.32% (2nd)	1.1% (4th)
Ho et al. (17)	Hong Kong	7393	0–14	Self-reported	1.79% (1st)	0.19% (8th)
Wu et al. (18)	Taiwan	30,018	All	Self-reported	7.23% (1st)	1.32% (2nd)

Figure 7.1. *Prevalence of seafood allergy.*

Source: Wai CYY, Leung NYH, Leung ASY, Wong GWK and Leung TF (2021) Seafood Allergy in Asia: Geographical Specificity and Beyond. Front. Allergy 2:676903. doi: 10.3389/falgy.2021.676903.

Another problem concerning the allergenicity of seafood is the high number of species that are consumed, giving rise to an extensive number of protein profiles involved. Furthermore, cross-reactivity between the seafood species and various other allergenic sources may complicate the situation further for those who are sensitized to seafood. An understanding of mechanisms involved in the allergic responses helps improve diagnostic approaches and survival methods that are necessary to improve the quality of life of affected allergic individuals.

With advances in research, knowledge improves in relation to seafood allergens and their health effects. This chapter shall go deep into the details of the allergenicity of seafood, which shall be elaborated on with respect to the major allergenic proteins, the immune biology mechanisms, and the prevalence of such allergies. It also provides the latest trends in diagnosis and management techniques, possible advancement in the regulatory framework, and new emerging trends in this field.

7.1. COMMON SEAFOOD ALLERGENS

Globally, seafood allergies rank among the most common food allergies, they still affect a fair percentage of the population. These allergies are caused by different proteins present in both fish and shellfish. Knowing the major allergenic proteins in seafood and the prevalence of the allergy is hence of essence in management and risk mitigation measures on seafood consumption.

7.1.1. Major Allergenic Proteins

The major allergens in seafood are proteins which have the potential of causing an immune response in sensitive individuals. These kinds of proteins vary from fish to shellfish and even among different species of these products.

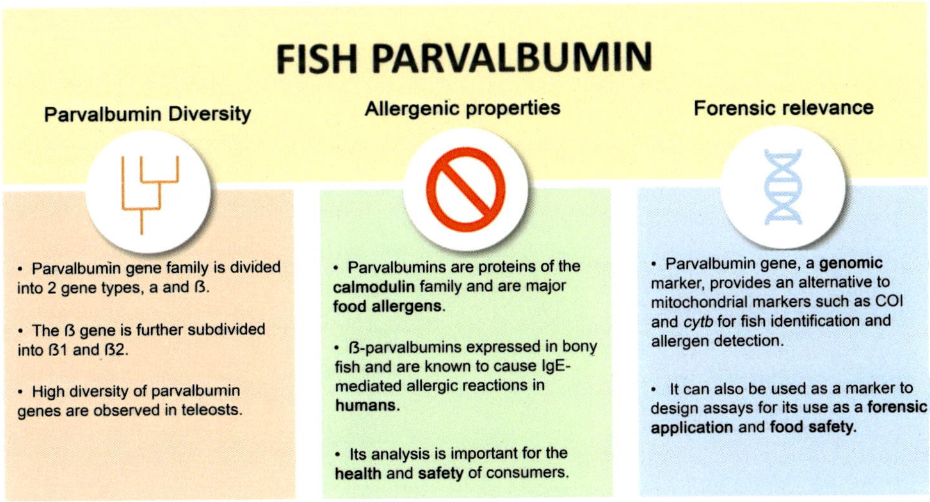

Figure 7.2. *Overview of the importance and applicability of fish parvalbumin.*

Source: Mukherjee S, Horka P, Zdenkova K, Cermakova E. Parvalbumin: A Major Fish Allergen and a Forensically Relevant Marker. Genes. *2023; 14(1):223. https://doi.org/10.3390/genes14010223.*

The major allergenic protein of fish is parvalbumin, a calcium-binding protein in muscle tissue. Parvalbumin is extremely resistant to heat and digestive enzymes, an ideal allergen. This protein is highly present in many of the most frequently consumed fish, for example, cod, salmon, and tuna. It accounts for the majority of allergic reactions against finned fish (Figure 7.2).

The allergens in shellfish are highly varied and represent a large group of proteins. The most important allergen in this group of crustaceans, which includes shrimp, crab, and lobster, appears to be tropomyosin, a muscular contraction protein. Tropomyosin is an important allergen in many individuals and has the curious property of being heat-stable and acid-stable and may become even more antigenic as digestion fragments. The other important allergens identified in crustaceans are arginine kinase and myosin light chain (Figure 7.3).

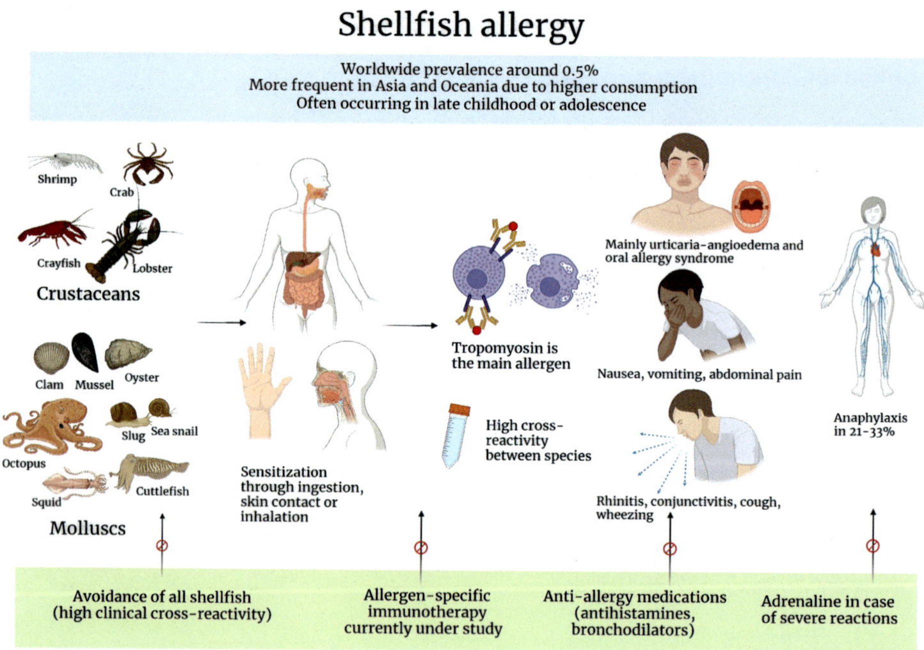

Figure 7.3. *Shellfish allergy.*

Source: Giovannini M, Beken B, Buyuktiryaki B, Barni S, Liccioli G, Sarti L, Lodi L, Pontone M, Bartha I, Mori F, et al. IgE-Mediated Shellfish Allergy in Children. Nutrients. *2023; 15(12):2714. https://doi.org/10.3390/nu15122714.*

The species involved with the allergenic proteins differ vastly and have no constant in mollusks; they range across clams, oysters, squid, and scallops. These various proteins induce allergic reactions through different mechanisms, although tropomyosin is said to be among the major allergens in mollusks, while others include hemocyanin and sarcoplasmic calcium-binding protein. Such structural properties on these proteins make them allergenic and thus cross-reactive. Cross-reactivity is an allergic reaction to more than one type of seafood because of almost the same protein structure.

7.1.2. Prevalence of Seafood Allergies

Seafood allergy is quite prevalent and can occur at any point in one's life; however, it is more common in adults as compared to children. Geographically, seafood allergy prevails more in those regions where seafood forms a staple part of the diet.

In the United States, seafood allergy accounts for about 2–3% of the adult population and 0.6% of children (Sicherer, Muñoz-Furlong, and Sampson, 2004). They are quite common in patients with other allergic diseases such as asthma or allergic rhinitis. Fish allergy is relatively less frequent compared to shellfish allergy. Crustaceans account for a higher percentage of cases of shellfish allergy compared to mollusks.

In Europe, the prevalence of seafood allergies is as huge and almost the same as that reported in the United States. In Asian countries, where seafood intake is usually higher, the prevalence of seafood allergy can be even greater. For example, in Japan, some studies have documented a higher rate of seafood allergy compared with Western countries. While in a few African and Latin American countries, the prevalence of seafood allergy is not very well documented, it is likely to be influenced by dietary as well as genetic factors.

The reasons for the geographical differences in the prevalence of seafood allergies are multilevel. Dietary exposure is a big factor—populations consuming more seafood are naturally at higher risk for developing allergies to these foods. There are also genetic predispositions, environmental factors, and the allergenicity variable of the different seafood species adding to these differences.

Seafood allergies may cause a very wide range of symptoms, ranging from mild to severe. Some of the common manifestations include skin reactions like hives and eczema, gastrointestinal symptoms such as nausea, vomiting, diarrhea, and respiratory symptoms such as wheezing, coughing, and nasal congestion. More severe cases, however, are those in which seafood allergies trigger anaphylaxis, a life-threatening reaction in individuals characterized by difficulty breathing, a drop in blood pressure, and loss of consciousness. Anaphylaxis needs immediate medical attention with the administration of epinephrine. The burden of seafood allergies extends beyond the physical symptoms to the quality of life of the patient. For those suffering from an allergy, avoidance of seafood is obligatory, which can pose significant difficulties since seafood is used in almost every cuisine and owing to issues related to cross-contamination during food preparation. This requires constant vigilance, which can result in anxiety and social limitations because eating out and traveling can become so complicated.

Diagnoses of seafood allergies typically include medical history, skin prick tests, blood tests measuring specific IgE antibodies, and oral food challenges done under medical supervision. An accurate diagnosis is critical in avoiding these foods effectively.

7.2. MECHANISMS OF SEAFOOD ALLERGENICITY

Seafood allergenicity encompasses rather multifaceted mechanisms of immune response and, hence, is affected by a set of determining factors in terms of severity or prevalence of allergic reactions. Understanding these mechanisms and influencing factors would be of basic importance in creating strategies for the proper management and prevention regarding seafood allergies.

Immunoglobulin E (IgE) antibodies are responsible for the major immune response to seafood allergens. The immune system identifies some specific proteins from seafood as harmful external invaders in the case of individuals suffering from the allergy. The process takes place in the following ways:

- **Sensitization:** This is the initial phase of an allergic attack. In this step, the immune system gets exposed to an allergenic protein for the first instance. In a seafood allergy, this could occur from ingestion, inhalation, or dermal exposure. Antigen presentation causes the immune system to perceive the protein as a threat, leading to the eventual initiation of a response involving the production of certain IgE antibodies against the antigen. These IgE antibodies coat the surface of mast cells and basophils, which are both types of white blood cells with important roles in allergy.
- **Re-exposure and Activation:** On re-exposure to the same allergenic protein, the allergen binds to the IgE antibodies on the mast cells and basophils. Such bind activates said cells, allowing their degranulation. So activated, these cells degranulate in the surrounding tissues and release a variety of chemical mediators made up of histamine, leukotrienes, and cytokines.
- **The Inflammatory Response:** Mediators, especially histamine, will elicit all the symptoms of allergic reactions. Histamine can increase the permeability of blood vessels to produce swelling, redness, and

itching. It also inflicts the constriction of smooth muscles leading to wheezing and breathing difficulty. Other mediators bring about the inflammatory response, worsening the symptoms.

Specific symptoms and degree of severity vary from person to person and by the extent of exposure to the specific allergen. Mild ones will manifest with rashes on the skin and gastrointestinal discomfort; serious reactions will cause anaphylaxis, a potentially fatal condition, which is a reason for immediate and urgent medical attention.

7.2.1. Factors Influencing Allergenicity

- **Protein Structure and Stability:** Allergenic proteins with the potential to induce allergy exhibit an important factor in terms of their structure and stability. More specifically, the proteins that are more resistant to heat and digestive enzymes are those that can more probably induce allergy (Fu et al., 2019). For example, parvalbumin from fish and tropomyosin from shellfish are very stable proteins, easily surviving cooking and digestion without losing any of their allergenic properties. It is precisely because of this that they are resistant to degradation and can therefore withstand and reach the immune system in their entirety, thereby increasing the possibilities for an allergic response.
- **Genetic Predisposition:** Individual susceptibility to seafood allergies is influenced by genetic factors. One inherits the risk of seafood allergy from one's family, especially if there is a history of allergies, asthma, or atopic conditions. Genetic predisposition does affect how much IgE antibody an immune system is likely to produce in response to specific allergens.
- **Age and Immune System Maturity:** The second influential cause of an allergy is age. It has been documented that seafood allergies are more common among adults compared to children, most likely due to differences in the maturity of a person's immune system and patterns of exposure. However, the majority of children with seafood allergies do carry them over into adulthood, and new allergies can develop at any time in a person's life.
- **Environmental Factors:** They including pollutants and changes in lifestyles, are very important to the changeability and seriousness of seafood allergies. Pollutants such as diesel particles and tobacco smoke have been known to increase allergic sensitization due to increased permeability across mucosal barriers and the promotion of inflammation.

- **Cross-Reactivity:** Another factor that might impact seafood allergy manifestations is cross-reactivity between different allergens. Cross-reactivity refers to the recognition of similar protein structures by the immune system from related species, causing an allergic reaction to the particular allergens. For instance, patients allergic to shellfish may react to dust mites or other invertebrates due to the common presence of tropomyosin.
- **Food Processing and Preparation:** The processing and preparation methods may either increase or decrease the allergenicity of seafood. Certain processes, such as boiling and frying, can change the structure of some allergenic proteins and either diminish or enhance their allergenic properties. Sometimes, heat treatment can denature the proteins, reducing their binding capacity to IgE antibodies, and the immune response could be less severe. However, in highly stable proteins like parvalbumin and tropomyosin, cooking is likely to have little or no effect on lowering allergenicity.
- **Gut Health and Microbiota:** The gastrointestinal tract's health and gut microbiota constitution can modulate the body's immune response against allergens. A healthy gut barrier and a balanced constitutive microbiota are believed to modulate the immune system and reduce the risk of sensitization to food allergens. Gut health disturbances through antibiotic use, infections, and diet might raise the chances of acquiring food allergies, including seafood allergies.
- **Cumulative Exposure:** Repeated exposure to seafood allergens over time can be an additional factor in the intensity of allergic reactions. Sensitization and increased potential risk of having allergic responses during following exposures can result from cumulative exposure. This is particularly applicable to people who frequently consume seafood or who work in an environment with high sea food allergen concentration, such as seafood processing plants.

7.3. DIAGNOSIS OF SEAFOOD ALLERGIES

The diagnosis of seafood allergies requires a mix of clinical appraisal, patient history, and some specific testing methods for the exact identification of allergens that cause allergic reactions. Because there exist a very great number of species within seafood and the probability of cross-reactivity among them, the process is hence very complicated.

7.3.1. Testing Methods and Protocols

- **Patient History and Clinical Evaluation:** The diagnostic procedure starts with a view of a broad patient history and clinical examination. What the physicians try to gather from the information includes the symptoms, the period the reactions occur, and the possible causes such as types of seafood eaten, the process of preparation, and other probable allergens. The medical history of the patient, even concerning the family, of allergies or atopic conditions is also looked into. This is a preliminary assessment to figure out the possibility of having a seafood allergy and direct further testing.
- **Skin Prick Tests (SPT):** Usually, seafood allergics are diagnosed using skin prick tests. Allergen extracts are placed on the skin of the forearm or back using a series of tiny pricks or punctures. The SPT will be positive when the patient is allergic to a seafood protein, and a local reaction, like redness and swelling, which is officially called a wheal and flare reaction, will occur at the test site. SPTs are relatively fast and cheap, with immediate results. But they carry false results and should be used in conjunction with a mixture of other procedures.
- **Serum-Specific IgE Testing:** This test involves blood samples that are drawn to detect infections inside the blood samples that could reveal allergens to seafood. To this date, the predominantly applied practice is the use of ImmunoCAP in the measurement of specific IgE antibodies in serum and the enzyme-linked immunosorbent assay. These can identify allergenic proteins in seafood and provide quantitative response data concerning IgE. The test is less invasive compared to the skin prick test, but it also happens to be more expensive and results may take longer. Although the clinical relevance to patient history is important for a correct diagnosis, high levels of specific IgE antibodies reflect a greater risk of triggering an allergic reaction.
- **Oral Food Challenge (OFC):** It is generally accepted as the diagnostic gold standard for food allergies-including seafood allergies. This test is done by letting the patient ingest predefinable increasing doses of the test food under close medical supervision. This test is performed only in the monitored place, such as a clinic or hospital, where one can receive emergency treatment after a severe reaction. The OFC gives definitive confirmation of allergy, but normally due to the danger of severe reactions, it is used as a method of last resort when other testing leaves unclear results.

- **Component-Resolved Diagnostics (CRD):** Component-resolved diagnostics is an advanced method used to detect particular allergenic proteins within a food allergen (Calamelli et al., 2019). Instead of whole food allergen extracts, CRD tests for constituent protein components, which give much more specific information about the specific proteins involved in provoking the allergy. It can differentiate primary sensitization from cross-reactivity and thus offers more information for the detailed understanding of the allergy profile of the patient. CRD is mostly helpful in extended and intricate situations with more than a single allergy to food or suspicion regarding cross-reactivity. In total, the detection of allergen causes in the case of seafood allergies depends on a careful clinical evaluation combined with detailed patient history and testing according to the species, which is considerable. A diagnosis is crucial for the management and prevention of allergic reactions, improving the outcome of a patient, and also guiding the recommendations that are to be made for dietary management.

7.3.2. Challenges in Diagnosis

The unique challenge to the diagnosis of seafood allergy lies in cross-reactivity with other species of seafood and allergens. The above scenario shows how tropomyosin protein is found in shellfish and mites and the protein frequently causes symptoms in sensitized individuals to such allergens. This presents some false positives and hence complicates the diagnostic process. Distinguishing real food allergy from another disease would be possible only if the patient had properly reviewed the history and if advanced diagnostics, such as CRD, were used.

Allergenic variability between different seafood items and even within the same species could comprise many factors, including geographical origin, preparation, and environmental conditions. Such variability complicates efforts to standardize allergen extracts used in testing and makes the result inconsistent. In addition, some patients might be allergic to only certain kinds of seafood and hence need a diagnosis performed for several species to get the correct diagnosis.

False positives and negatives: In both skin prick and serum-specific IgE tests, false-positive or false-negative results can be obtained. Due to nonspecific binding or cross-reactivity with other allergens, false-positive results are very likely and will generate unnecessary dietary restrictions. False-negative results can occur due to the low affinity of IgE antibodies or lack of allergenic protein in the test extract. These limitations highlight the importance for test results to be correlated with clinical history and for additional testing methods, such as the oral food challenge, to be considered.

Anxiety and Patient Risk: The diagnostic process, especially tests that include provocative action by the allergens, can provoke serious anxiety in the patients. For the risk of serious allergic reactions from testing, especially oral food challenges, proper considerations and precautions are to be taken. This is so crucial in the diagnostic process, especially in ensuring that patients are comfortable and safe psychologically.

Variation in Protocols and Allergen Extracts The lack of standardization of protocols and allergen extracts causes variability in test results and in the diagnostic accuracy of seafood allergies. The preparation of allergen extracts for skin prick tests or serum-specific IgE tests under different conditions may induce changes in sensitivity and specificity. Efforts are ongoing to standardize diagnostic methods and the quality of allergen extracts, but they continue to pose a problem.

7.4. MANAGEMENT OF SEAFOOD ALLERGIES

An approach to the management of seafood allergies needs to be based on proactive and reactive strategies. While strict dietary avoidance is the mainstay of management, the importance of emergency management plans cannot be undermined for those unexpected accidental exposures.

Dietary avoidance forms the mainstay for managing seafood allergy. Those with a seafood allergy should avoid the allergenic seafood in its entirety, whether fish, shellfish, or groups of these like crustaceans or mollusks. Basically, it involves being very attentive to food labels, asking the right questions when eating out, and being cautious of cross-contamination that may occur during food preparation. Seafood can often be hidden in dishes, sauces, or processed foods, so it's very important for allergic individuals to become knowledgeable with regard to ingredients and food preparation practices.

Dietary avoidance needs education and awareness. Allergic individuals and their families should be knowledgeable regarding seafood products and by-products that may be a potential risk. This includes the meaning of terms like "fish oil," "anchovy paste," and "seafood broth," which may be present on an ingredient list. Moreover, they have to focus not only on the obvious source but also on certain flavorings and supplements. Those with allergies should inform this allergy to restaurant people and any food providers so that food can be prepared safely for them.

However, accidental exposures can still happen despite best efforts for avoidance. Therefore, individuals with seafood allergies must be adequately prepared to handle allergic reactions promptly. The best approach to management in an emergency is to carry an epinephrine auto-injector at all times. Epinephrine represents the first line of treatment during anaphylaxis, which is a serious and potentially life-threatening allergic reaction. If administered promptly, epinephrine reverses the symptoms of anaphylaxis and perhaps avoids disastrous complications. Education regarding the proper administration of epinephrine autoinjectors is essential for allergic patients and their caregivers. Awareness regarding when to use i.e., during severe allergic reactions, and how to administer the epinephrine will save lives. Beyond just carrying an auto-injector on their person, an emergency action plan should include calling 911 after administering epinephrine, as some patients may need further medical evaluation and treatment after the injection.

Other ways in which antihistamines might feature in an emergency kit include the treatment of milder allergic reactions, such as urticaria or pruritus. In no way, however, can they replace epinephrine if a patient has a severe reaction. Corticosteroids may also be indicated to decrease inflammation after an allergic reaction but usually as adjunctive rather than primary therapy.

Long-term care for seafood allergy also involves frequent follow-up care with the health provider, including an allergist if necessary. Periodic assessments will update the current condition regarding the allergy process, care for related diseases such as asthma or eczema, and revamp emergency treatment plans as required. In some cases, food challenges may be carried out under controlled conditions by some allergists to determine if the allergy has persisted or if desensitization is possible.

Avoidance strategies must also reach into the home environment by ensuring that cooking utensils, cutting boards, and surfaces are cleaned thoroughly to avoid cross-contact with seafood allergens. The education on the importance of preventing cross-contamination shall be conveyed to the family members or caregivers, as well as the understanding of how to recognize signs of an allergic reaction. Awareness and advocacy are also part of management for seafood allergy. A support group and advocacy organization can help one get emotional support and the latest details about living with an allergy. Such a group may facilitate better food labeling and policies that protect people with food allergies at schools, workplaces, etc.

7.5. CROSS-REACTIVITY IN SEAFOOD ALLERGIES

Cross-reactivity is a condition where the immune system will respond not only to a primary allergen but also to the proteins of other, usually related, substances due to structural similarities. In this way, some people are allergic to one group of seafood and react to others, or even to nonseafood allergens. Knowing the types and implications of cross-reactivity is important in the management and mitigation of seafood allergy risks.

One of the primary mechanisms underlying this phenomenon is the existence of homologous proteins between species. Major allergenic proteins, such as parvalbumin in fish and tropomyosin in shellfish, have already been shown to be conserved in a large proportion of species (Ruethers et al., 2018). For example, a person allergic to parvalbumin in cod may cross-react to the exact proteins in salmon or tuna. A person allergic to tropomyosin in shrimp may react also to crab, lobster, or even mollusks like squid or octopus. This complicates avoidance of all possible allergens because the individual may have to avoid many different types of seafood.

Cross-reactivity, however, is not limited to seafood alone. Sometimes, individuals allergic to seafood tend to react to other allergens as well. For example, tropomyosin is present not only in shellfish but also in insects and dust mites. Therefore, individuals who are allergic to shellfish will also react to these antigens due to the similarities in their protein structure. This is pan-allergen cross-reactivity, where one allergen protein in multiple sources invokes similar immune responses.

Cross-reactivity in allergies can have significant implications for allergic individuals. First, there is the added complexity of diagnosis and management. A person may ingest another related allergen and unconsciously experience an allergic reaction; thus, identification of all possible allergens must be exact. This involves comprehensive allergy testing and evaluation, including component-resolved diagnostics for the identification of particular proteins to which a person is allergic. The conventional methods of skin prick tests and serum-specific IgE may sometimes be indefinite, especially in cases of cross-reactivity. There, advanced diagnostic tools are of great importance.

For those affected, this cross-reactivity implies heavier dietary restriction and increased vigilance. Greater restriction of a broader range of allergens puts one in the position of reading food labels more carefully, communicating with food providers more precisely, and having a greater overall depth of knowledge about cross-contaminants. Social settings and eating out can be particularly challenging, as cross-contamination within kitchens can easily occur. Reactions can result from traces of the allergen, so utmost attention to food preparation and sourcing must be taken.

This also has implications for the quality of life: strict diet restriction and constant vigilance may mean social and psychological stress. These persons could fear going out to eat, traveling, or doing any other food-related activity associated with social events. Continuous stress because of this factor can affect their general well-being, underlining their need for support systems and resources in coping with multiple food allergies.

Moreover, this cross-reactivity can have implications for the way education and advocacy in issues of allergy are conducted. In this respect, awareness of cross-reactivity assumes a significant role in making allergic environments safer. It means advocating for clearer food labeling, better food safety handling in restaurants and processing plants, and greater education among the general public about the risks involved with cross-contamination.

Cross-reactivity knowledge at basic research and clinical practice levels is important in improving treatment strategies and management techniques. Investigators are pursuing a mechanism of desensitizing the person to multiple allergens simultaneously, which brings new hope to those affected by cross-reactive allergies. That may eventually be possible with the development of immunotherapy and individual approaches to care.

7.6. PREVENTION OF SEAFOOD ALLERGIES

The prevention of seafood allergy requires a multisectoral approach including effective allergen detection and labeling practices on the one hand, and comprehensive consumer education and awareness programs on the other.

7.6.1. Allergen Detection and Labeling

The detection of allergens in seafood and seafood products is an important step in preventing allergic reactions. It involves a stringent test of food products to determine the presence of allergenic proteins (Dong and Raghavan, 2022). The techniques applied in detecting allergens include the enzyme-linked immunosorbent assay, mass spectrometry, and polymerase chain reaction techniques. These techniques identify and quantify certain specific allergenic proteins present in a food product with a view to proper labeling to ensure safety for consumption by allergic persons.

Probably one of the most used methods for the detection of allergens is ELISA. This technique identifies specific food proteins in food samples through antibodies. It is a very sensitive technique, thus able to detect traces of an allergen, which makes this tool very efficient in terms of food safety assurance. MS and PCR represent other useful techniques that provide good specificity and accuracy in the identification of their allergenic proteins and respective genetic markers (Figure 7.4).

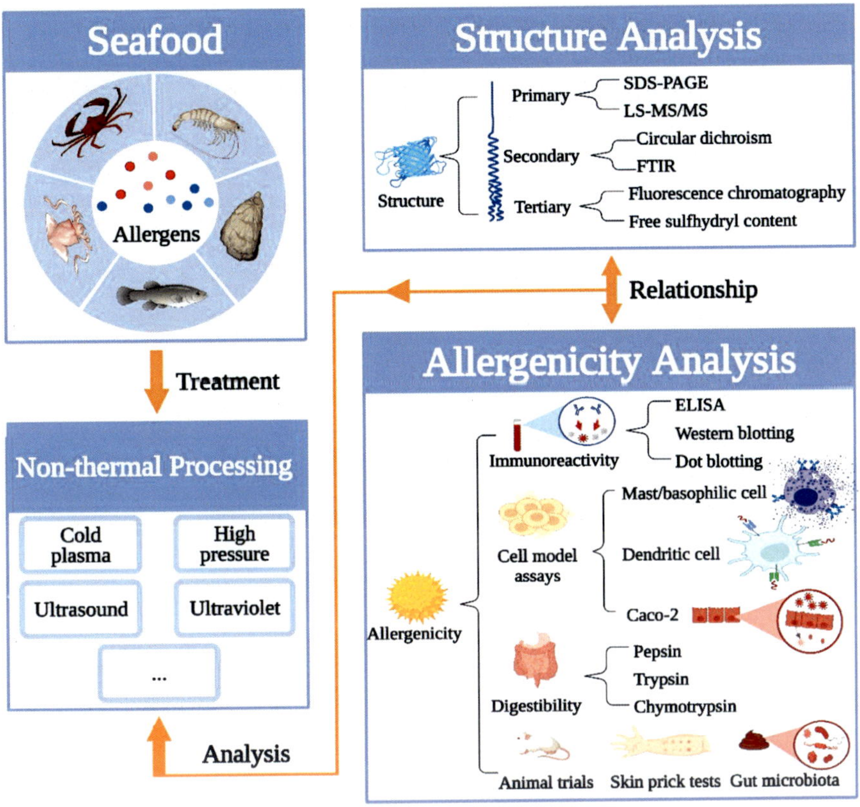

Figure 7.4. *Analysis of structure and allergenicity changes of seafood allergens induced by non-thermal processing.*

Source: Wang F, Zhong H, Cheng J-H. Comprehensive Analysis of the Structure and Allergenicity Changes of Seafood Allergens Induced by Non-Thermal Processing: A Review. Molecules. *2022; 27(18):5857. https://doi.org/10.3390/molecules27185857.*

The next important stage of prevention after the detection of allergens is accurate labeling. Country-wise, food labeling rules may differ, but the objective is to give consumers concise and complete information about the presence of allergens. The Food Allergen Labeling and Consumer Protection Act requires food products that contain any of the eight major allergens, which include fish and shellfish, to indicate the presence of these allergens on the label. Similarly, the European Union has legislation, under the Food Information for Consumers Regulation, requiring labeling for 14 allergens to include fish and crustaceans.

Accurate and clear labeling will help consumers who are allergic to a certain type of seafood avoid consuming products that have allergens. Ingredient labeling is important, with components that are allergenic highlighted or marked for easy readability and understanding. Added to this, precautionary labeling includes important information for the consumer on the risk of cross-contamination.

7.6.2. Consumer Education and Awareness

Consumer education on seafood allergies and ways of avoiding contact with allergens will become an important aspect of prevention in the future. Indeed, effective consumer education will have to include activities such as detailed information about the nature of seafood allergy, potential sources of allergens, and strategies for avoiding exposure.

One of the key issues to be addressed is consumer education regarding the variety of seafood and seafood-ingredient allergens. This would include less obvious sources of allergens, such as fish sauce, anchovy paste, and seafood extracts used in flavorings and broths. Consumers need to be educated to read food labels carefully and ask questions about ingredients when eating out.

It can come in many forms from many different sources, including health care providers, schools, community groups, and the Internet. Health care providers, especially allergists and dietitians, are on the front line for educating patients and their families about seafood allergy. They can advise on avoidance, management of symptoms, and preparation for an accidental exposure.

Workshops or educational seminars regarding food allergies could also be provided by schools and community organizations. These programs can serve to educate children and adults about the symptoms of an allergic reaction, give simple guidelines on the use of epinephrine autoinjectors, and teach the art of communication in monitoring allergies. Policies on allergy management should be developed in schools, which would include staff training to provide efficient and timely responses during allergic reactions and to make the environment safe for students with allergies.

Online resources and social media offer great potential means through which information on seafood allergy may reach the broadest possible audience. Websites focused on food allergies could provide education, recipes, and cooking hints for the safe handling of food. Social media campaigns will raise awareness and build a network where the person with food allergies feels supported.

Consumer education must focus on the importance of emergency preparedness. Consumers with seafood allergies should be able to identify early symptoms of an allergic reaction and respond quickly and effectively by carrying epinephrine auto-injectors and knowing the proper way to use them. This differs from the responsibilities that would extend to education for family members, friends, and caregivers on the management and emergency response toward such allergies.

Besides consumer education, the education of food industry personnel is very important in preventing allergic reactions. It involves educating chefs, restaurant staff, and food manufacturers about the danger of cross-contamination and adopting practices to minimize it through the use of separate utensils, separation of areas for allergen-free food preparation, proper cleaning of the equipment, and communication with customers regarding the ingredient list and possible allergens.

7.7. REGULATIONS REGARDING SEAFOOD ALLERGENS

The regulations around seafood allergens become very important in ensuring the protection of consumers with allergies to seafood and food safety. Within these regulations are comprehensive frameworks of labeling requirements that ensure persons at risk of allergic reactions from these foods are informed and protected. Country-specific regulations may differ, but international standards allow for international harmonization of efforts toward coherent management of allergen risks.

The Food Allergen Labeling and Consumer Protection Act, passed by Congress in 2004, requires food manufacturers in the United States to clearly identify any products that contain major food allergens (Fortin, 2006). Fish and crustacean shellfish are two of the eight major allergens classified by this act. FALCPA enforces that on the product label, the specific type of fish and shellfish be identified to allow consumers to recognize the presence of a food allergen. This act has played a very important role in the reduction of risk associated with accidental exposure to food allergens obtained from seafood and in improving the safety of food products in general.

Under the European Union's Food Information for Consumers Regulation, food products are obligated to declare the presence of any of the 14 recognized allergens, one of which is fish and crustaceans. This means that an ingredient that forms an allergenic constituent must be emphasized in the ingredients list, usually using the resources of bold, italic, or underlined text. As such, this requirement enables consumers to readily identify possible allergenic constituents of the food product for their safety. Moreover, it has also made a stipulation for the EU that non-prepacked foods i.e., which are foods sold in restaurants and catering, shall also indicate allergen information, hence guaranteeing comprehensive protection across varied contexts of food consumption.

At the international level, the Codex Alimentarius Commission is a joint body of the Food and Agriculture Organization and the World Health Organization that elaborates standards for food across the world. Based on the Codex guidelines, major allergens (into which category fish and shellfish fall) must be declared on food labels to protect allergic consumers. This guideline acts as a benchmark for many countries that are developing or updating their food allergen regulation.

Regulatory frameworks also regulate the issue of cross-contamination, which can be a major concern for people with seafood allergies. During food processing, preparation, or packaging, there is a risk of accidental exposure of an allergen to a non-allergenic food item, causing cross-contamination. Many regulatory bodies, like FDA and EFSA, provide guidelines on GMPs to help safeguard against this risk factor. These practices include thorough cleaning of the equipment, separate utensils for both allergen and non-allergen foods, as well as strict hygiene measures to avoid cross-contact.

Labeling and manufacturing practices aside, some countries are developing precautionary labeling, such as "may contain" statements, to further help consumers. This particular label generally denotes that the product may have had contact with an allergen while being produced, even though the allergen is not one of the intended ingredients. While precautionary labeling is aimed at further protecting consumers, several concerns arise, one of which is the potential to cause unnecessary dietary restrictions for allergic consumers. Work is currently underway to increase the validity and reliability of these statements so that they are actually risk-based rather than precautionary.

The other important element in these frameworks is enforcement in the area of allergen labeling. Not only do these regulatory agencies have routine inspections, but they also oversee food products to verify compliance with provisions on allergen labeling. In case of non-compliance, a company would incur penalties such as product withdrawal and fines. Such enforcement sustains very tight food safety standards, thus keeping the manufacturers responsible for providing protection for consumers with food allergies.

Consumer education and awareness are intrinsic to the effectiveness of allergen regulations. Indeed, regulatory bodies will often join forces with public health organizations and advocacy groups in consumer education to label reading, recognizing allergen information, and the need for allergen management. Education efforts empower persons with seafood allergies to make appropriate decisions and to be their own advocates.

This does not mean that there are no current issues or areas of improvement with these robust frameworks. Indeed, at times, this variability in the requirements for labeling allergens becomes confusing to international travelers or even hinders global food trade. Harmonization efforts concerning standards on labeling allergens, like those moved forward by organizations such as Codex Alimentarius, would then be key to ironing out these inconsistencies and guaranteeing protection at a universal level for allergic consumers.

7.8. EMERGING TRENDS IN SEAFOOD ALLERGENICITY

The emergence of novel allergen identification is the major trend in seafood allergenicity. Several allergenic proteins from different seafood species have already been known, such as parvalbumin from fish and tropomyosin from crustaceans. Recent examples of novel allergenic proteins from the seafood species include the allergens identified in this work: enolase and aldolase from shellfish. These new allergens extend the current view of the allergenic potential of seafood and underline the need for comprehensive allergen profiling.

Proteomics and genomic approaches, combined with mass-spectrometry technologies in peptide and protein sequencing, have added much to the identification of novel allergenic sources. Proteomic approaches enable the detailed structural analysis of proteins and thus direct the identification of which epitopes cause allergenic reactions. Genomic sequencing provides the genetic basis of allergenicity, enabling the identification of new allergens more effectively. These changing technologies in the field are offering better illumination of the molecular mechanisms that underlie seafood allergies.

Cooperativism represents a fast-growing area in hypoallergenic seafoods. Current research is on the development of seafood that still maintains its nutritional and sensory value, in regard to the reduced allergenic potential. One possible approach involves the genetic modification of allergenic proteins that find themselves in seafood. Altering the structure of a protein may also lead to a reduction of binding sites for anti-IgE antibodies and a reduction in allergenicity. For instance, another product that is currently in the works is the potential to utilize CRISPR-Cas9 technology in the gene-editing of allergenic proteins, particularly in fish and shellfish.

Such techniques of processing food that reduce the allergenic proteins or partly denature them to convert the structure of the proteins into possibly less antigenic forms may include thermal processing, enzymatic treatment, or fermentation. For instance, studies have confirmed that enzymatic hydrolysis degrades allergenic proteins into small peptides, most of which do not elicit immune responses. This process is under optimization to make sure that the final product is safe and palatable for consumers.

Another approach to developing hypoallergenic seafood products is through the use of novel ingredients and formulation. Researchers have also been looking into nutraceutical-based seafood analog production approaches with the use of plant and insect protein sources that hold promise with the absence of the risk of allergic reactions. The growing trend in the direction of natural and alternative proteins, such as plant-based, holds strong potential for those individuals who are sensitive or allergic to seafood to obtain similar nutrients without the risk of becoming allergenic.

Component-resolved diagnosis is an example of another emerging trend by the improved diagnostic tools for the diagnosis and management of seafood allergies. It is a diagnostic tool used to identify specific allergenic proteins in the food. The technique actually gives more explicit information on the allergen profile and thus allows a clinician to make management strategies in an individualized manner. It can aid in the differentiation between primary sensitization to specific seafood proteins and cross-reactivity with other allergens, thus achieving diagnostic precision to provide suitable dietary advice.

Immunotherapeutic approaches are now being discussed for the treatment of seafood allergies. One such approach is oral immunotherapy (OIT), in which some degree of tolerance is developed upon gradual exposure to small quantities of the allergen. OIT has given some promising results concerning peanut and milk allergies, but with seafood allergies, it has only been used in the preliminary study. However, even these preliminary studies show that immunotherapy might be held in patients with seafood allergies to desensitize them, thereby reducing the severity of the reactions and improving quality of life.

7.9. FUTURE DIRECTIONS IN SEAFOOD ALLERGEN RESEARCH

Allergen identification and characterization remain a major ongoing area of novel research. State-of-the-art proteomics and genomics methods are being applied in the discovery of new allergenic proteins in a wide range of seafood species to build up this complete allergen profile. This will further enhance diagnostic accuracy and better allergic management strategies.

The other important issue will be to improve the diagnostic accuracy. The evolving field of component-resolved diagnostics is about to revolutionize allergy diagnostics by giving an accurate, specific identification of allergenic proteins. This will lead to better discrimination between primary sensitization and cross-reactivity, better adjudicating the dosages of therapies toward better personalization, and even improved diagnostic precision of diagnoses. Improved diagnostic tools will help in tailoring treatment options for every individual and making dietary recommendations.

As concerning potential therapeutic modalities, OIT and epicutaneous immunotherapy, or EPIT, are a very interesting idea, first, as far as their popularity is on the rise. The reason for their popularity is the developing desire to create tolerance in allergic subjects due to the graded administration of allergens. In fact, whereas OIT is so far reported to have successfully treated other food allergies, attempts for seafood allergies are under investigation. Preliminary results are encouraging for a decrease in the severity of allergic reactions and improved quality of life.

Another rather promising research area is the development of hypoallergenic seafood products in which allergenic proteins are modified or eliminated by molecular biological or food processing technology from seafood. Hypoallergenic seafood products try, by modifying the protein structure or using different protein sources, to develop seafood while maintaining their nutritional and sensorial quality but with a reduced allergenic potential.

Perhaps most importantly, there is a focus on public health initiatives and education. Increasing seafood allergy awareness, improving the quality of food labeling, and promoting safer food preparation will be key to lowering the allergy risk. Overall, the goal here is to have these interventions and improvements made in close collaboration.

Chapter 8

Biogenic Amines in Seafood

Contents

INTRODUCTION

Biogenic amines are a group of organic compounds produced through the decarboxylation of amino acids, most probably microbial in origin. These compounds, abundantly found in foodstuffs, including seafood, may have some roles in biological functions, but their accumulation in food products is very risky to human health. There are several conditions in which some biogenic amines like histamine, tyramine, and putrescine might form in seafood, most probably due to poor storage or handling conditions. Large amounts of these amines are related to food poisoning and allergic reactions, making them an important issue in food safety (Figure 8.1).

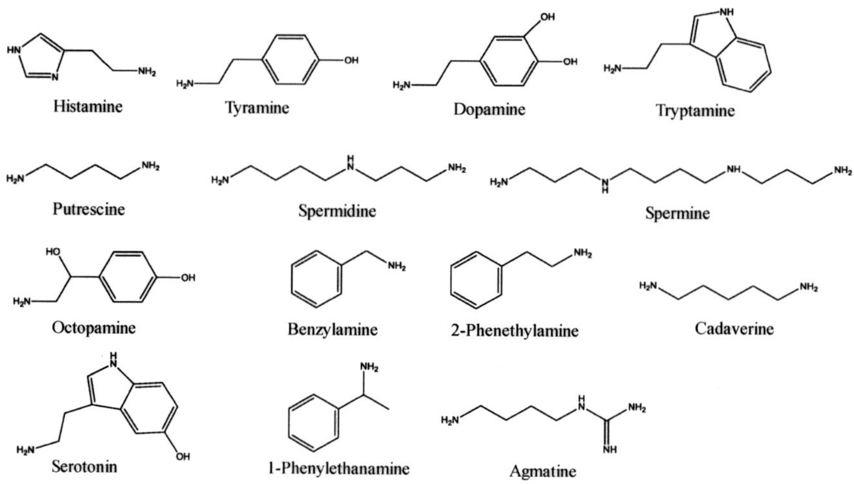

Figure 8.1. *Biogenic amine levels in food processing.*

Source: Li T, Wang R, Wang P. The Development of an Ultra-Performance Liquid Chromatography–Tandem Mass Spectrometry Method for Biogenic Amines in Fish Samples. Molecules. *2023; 28(1):184. https://doi.org/10.3390/molecules28010184.*

It is important that we understand how biogenic amines are formed, detected, and controlled so that we can guarantee seafood safety and quality. Improved analytical techniques have further developed the capability to monitor these compounds; however, maintaining safety levels is not easy throughout the whole supply chain. Legislation has thus set limits on permissible content levels of biogenic amines, spiraling further research in refining standards and more efficient mitigation strategies.

8.1. FORMATION OF BIOGENIC AMINES IN SEAFOOD

Biogenic amines are primarily organic nitrogen compounds formed by the microbial decarboxylation of amino acids. They are found in various foods, including seafood, and can, under certain conditions, attain large quantities, which can cause serious health risks when consumed in large amounts (Biji et al., 2016). Being able to understand the biochemical pathways involved in the formation of biogenic amines and the factors promoting their accumulation is essential in managing levels in seafood (Figure 8.2).

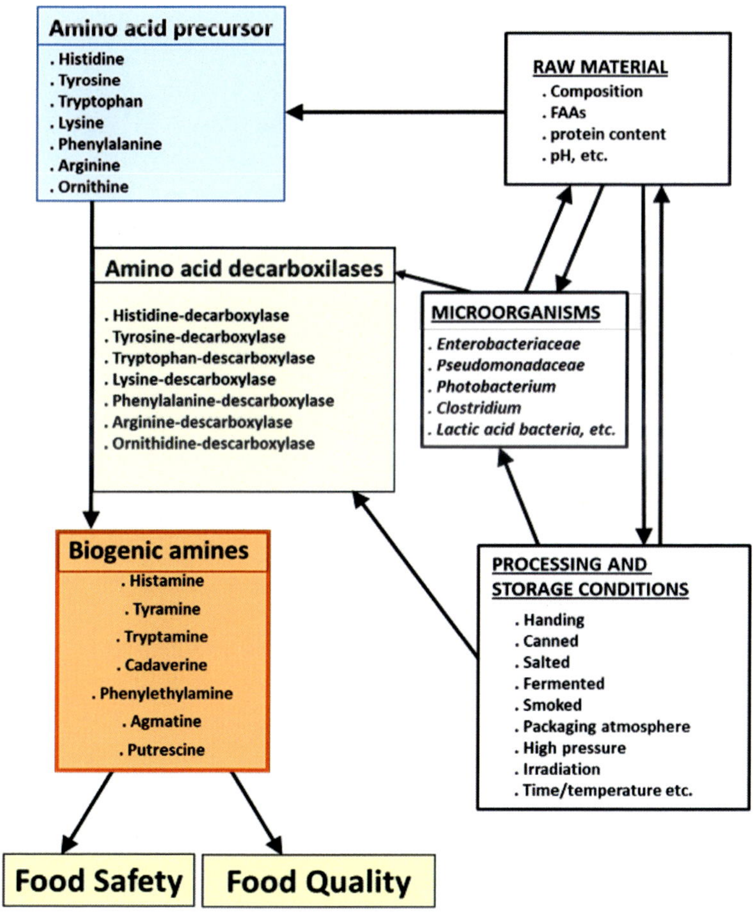

Figure 8.2. *Biogenic amines in meat and meat products.*

Source: Ruiz-Capillas C, Herrero AM. Impact of Biogenic Amines on Food Quality and Safety. Foods. 2019; 8(2):62. https://doi.org/10.3390/foods8020062.

8.1.1. Biochemical Pathways

Biogenic amines are majorly formed in seafood through decarboxylation of amino acids through microbial enzymatic means known as decarboxylases. These amines have different amino acid precursors:

- **Histamine Formation:** This results from the amino acid histidine due to the activity of the enzyme histidine decarboxylase. The above reaction is common in varieties of fish like tuna, mackerel, and sardines. The histidine content in these fish may be relatively high, which may form histamine, especially under conditions that promote the growth of relevant bacteria.
- **Tyramine Formation**: Tyramine is produced from tyrosine through the action of tyrosine decarboxylase. Tyramine is a constituent of many fermented foods and may also be present in seafood subjected to bacterial spoilage.
- **Putrescine and Cadaverine Formation:** These amines are the products of decarboxylation of ornithine and lysine in the presence of the enzymes ornithine decarboxylase and lysine decarboxylase, respectively. Putrescine and cadaverine have been used as indicators of spoilage, and the former potentiates the toxic effects of other biogenic amines such as histamine.
- **Spermidine and Spermine Formation:** These polyamines are formed from putrescine and act as significant growth factors in living cells. They can also be formed in spoiling seafood but are normally present in smaller amounts than histamine and tyramine.

8.1.2. Control Strategies

- **Temperature Control:** Keeping the fish at a low temperature below 4°C during storage and transit. This will slow down bacterial growth and enzyme activity. Quick cooling of fish following catching: deep-freezing must be done immediately; cold chain maintenance in processing and distribution.
- **Hygiene and Sanitation:** The process has very strict hygiene and sanitation procedures to reduce microbial contamination. This will involve frequent cleaning and disinfection of equipment, surfaces, storage areas, and good hygiene practices by the people handling the produce.
- **Preservatives:** The use of preservatives, such as organic acids, sulfites, or modified atmosphere packaging, allows for the prevention of the growth of bacteria and reduces the formation of amines. Such

methods are very useful for the extension of the shelf life of seafood while maintaining its safety and quality.

- **Monitoring and Testing:** The levels of biogenic amines in seafood products should be regularly monitored by using reliable methods of analysis. In this way, a problem can be noticed and controlled before it reaches consumers.

8.2. HEALTH RISKS ASSOCIATED WITH BIOGENIC AMINES

The accumulation of amines in high concentrations can pose serious health risks. Biogenic amines are related to serious health risks associated with their pharmacological activity at high concentrations. Of all these biogenic amines occurring in seafood, histamine, tyramine, putrescine, and cadaverine have been related to undesirable health effects when consumed in large amounts (Figure 8.3).

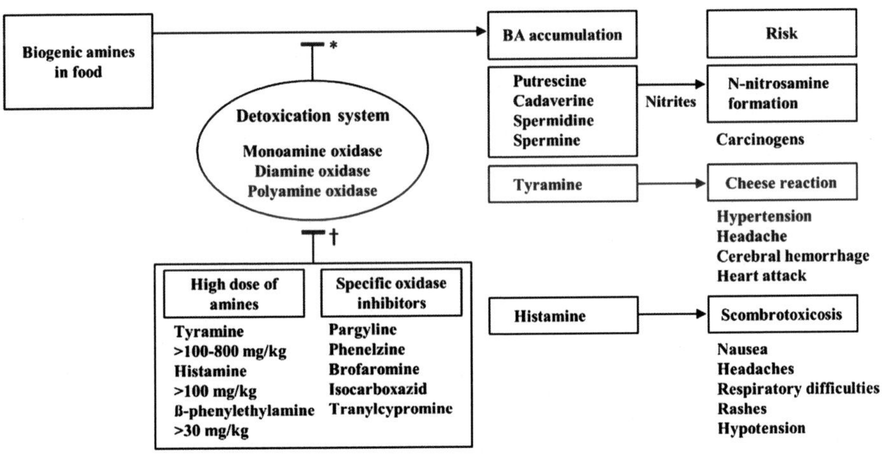

Figure 8.3. *Impact of biogenic amines.*

Source: Mah J-H, Park YK, Jin YH, Lee J-H, Hwang H-J. Bacterial Production and Control of Biogenic Amines in Asian Fermented Soybean Foods. Foods. 2019; 8(2):85. https://doi.org/10.3390/foods8020085.

- **Histamine:** It is the most known biogenic amine about foodborne illness, particularly that involving seafood (Prester, 2011). High levels of histamine can lead to scombroid poisoning, also known as histamine fish poisoning. This kind of poisoning is the result of the formation of histamine through the decarboxylation of histidine by bacteria in improperly stored fish. The symptoms and signs of scombroid poisoning are flushing, rash, headache, dizziness, palpitations, abdominal cramps, diarrhea, and vomiting. These symptoms usually start within minutes to hours after consumption and can range from very mild to very severe. In severe cases, histamine toxicity can lead to respiratory distress and cardiovascular complications.

- **Tyramine:** This is another biogenic amine that has immense effects on health, especially in those patients who are using MAOIs, a class of antidepressants. Tyramine can lead to hypertensive crises in such patients by releasing norepinephrine, which vastly increases blood pressure. Symptomatic tyramine toxicity includes severe headache, hypertension, palpitations, nausea, and vomiting. Still, high levels of tyramine may cause headaches and migraines even in subjects who do not use MAOI.
- **Putrescine and Cadaverine:** These are two diamines that have been used, together with the former, as indicators of spoilage and can potentiate other biogenic amines in their toxic effects. In themselves, putrescine and cadaverine are far less toxic, but can enhance the overall toxicity of seafood that has become contaminated by increasing intestinal permeability, allowing other harmful compounds to be more easily absorbed into the bloodstream. The symptoms associated with high levels of these amines include gastrointestinal disturbances, nausea, vomiting, diarrhea, and others.
- **Spermidine and Spermine:** While these polyamines have a role in carrying out cellular functions, they have also been related to negative health effects at high concentrations. They have been implicated in the promotion of growth of cancer cells and increasing risk for the development of cancer; however, their direct impact on human health at dietary levels is still under investigation.

8.2.1. Risk Assessment and Exposure Levels

The key sources of biogenic amines in seafood are microbial activity and improper storage conditions. Fish species with a high content of free amino acids, for example, histidine, are those highly predisposed to the formation of biogenic amines. These sources shed light on the understanding of implementing preventive measures to minimize contamination.

The concentration of biogenic amines in seafood may differ within a wide range due to the origin of seafood, its storage condition, and processing methods. With regard to food safety, regulatory agencies have set an allowable limit on biogenic amines. For example, the EFSA and FDA have set histamine limits for fish products. On the other hand, the FDA accepts 50 parts per million as the action level of histamine in fish. EFSA, however, has recommended a maximum limit of 200 mg/kg for fresh fish and 100 mg/kg for fishery products.

Tolerable daily intake is a measure that can be used in estimating a daily intake that, when taken for an entire lifetime, does not pose a significant health risk associated with the consumption of biogenic amine. The TDI value for histamine is usually considered to be 0.5 mg per kilogram of body weight per day. The above limit exposes individuals, especially sensitive ones, to the chances of toxicological effects.

With a view to compliance with the regulatory limits in force and consumer health protection, control of the levels of biogenic amines in seafood products is considered very important. Chromatography techniques, such as high-performance liquid chromatography coupled with mass spectrometry, are among the analytical methods for detecting and quantifying biogenic amines in food that are in use. Given this, using these techniques will give more accurate and reliable data about the levels of amines, thus helping to identify possible issues before products reach the market.

Good manufacturing practices, GMPs and Hazard Analysis and Critical Control Points, HACCP systems can prevent the formation of biogenic amines by ensuring proper storage temperatures, hygiene in handling and processing, and proper preservatives and packaging methods that may inhibit microbial growth. Training and education of seafood handlers and processors is also very important in minimizing contamination risks.

Consumer education on the risks associated with biogenic amines and on good seafood handling and storage practices will further reduce exposure levels. Consumers should be aware of the signs of seafood spoilage and not consume products that have become spoiled or have an unusual odor or color.

8.3. DETECTION OF BIOGENIC AMINES IN SEAFOOD

Biogenic amine detection in seafood plays a very vital part in the assurance of the quality and safety of foods. Therefore, it needs a highly accurate and reliable detection method for monitoring and controlling its concentration, as it is considered to pose health risks, especially at high concentrations. There are various analytical methods used in the detection of biogenic amines, all of which have their challenges, sensitivity, and accuracy.

High-performance liquid chromatography (HPLC) is one of the most applied methods for detecting biogenic amines in seafood. Based on its principle of operation, HPLC makes quantitative determinations with high sensitivity and specificity. The individual amines are separated on a chromatographic column, followed by detection with suitable detectors like UV or fluorescence detectors. HPLC allows the identification and quantitation of several amines in one analysis and, therefore, is very useful work requiring overall composition determination. However, this method involves the use of very expensive equipment, well-trained personnel, and complex sample preparation, which is time-consuming and expensive.

The other technique used a lot in the gas chromatography, is normally combined with mass spectrometry and abbreviated as GC-MS. GC-MS is a very sensitive tool that could include molecular structure details of biogenic amines, increasing therefore the accuracy of its identification. This technique is especially applied in the detection of volatile amines. However, GC-MS also involves complicated procedures of sample preparation and derivatization, which may add variability and become a possible source of error. In spite of all the complications, GC-MS is still an effective method of detecting and analyzing biogenic amines.

Another easier and faster technique for the detection of certain biogenic amines is the enzyme-linked immunosorbent assay. In this case, ELISA makes use of antibodies to bind the specific amines and produce some measurable signal proportional to the concentration of the amine being measured. The method is relatively easy to use and quick; thus, it can easily be used in routine screening. However, ELISA methods are less comprehensive compared to chromatographic techniques since they usually determine only one or a few amines at a time. Moreover, the specificity and sensitivity of ELISA depend on the used antibodies, which, in turn, differ in quality.

Biosensors are the developing technology for the determination of biogenic amines. The devices make use of biological molecules like enzymes or antibodies that bind selectively to the amines in question and are linked to an appropriate detector producing an electronic signal (Vasconcelos et al., 2021). Therefore, the advantages are that biosensors provide fast, real-time detection and can be designed for on-site testing, which means immediate results. They are relatively easy to use and can be cost-effective for routine monitoring. The biosensors for biogenic amines are under development and validation; their sensitivity and specificity have not yet been confirmed as reliable.

Each of these techniques comes with problems that may influence the correctness and reliability of the determination of biogenic amines. However, all analytical techniques rely on correct sample preparation; inappropriately treated samples could be degraded or contaminated, which again may provide wrong measures. A very important factor could be the complexity of seafood matrices containing a variety of proteins, lipids, and other compounds interfering with the detection process in such a way that biogenic amines may not easily be isolated and quantified.

Calibration and standardization are very important to obtain the most accurate results, which are comparable among various laboratories and studies. The use of standard reference materials and calibration graphs thus yields less variability and more consistency of measurement. Moreover, analytical methods should be regularly validated with relevant quality control procedures if their reliability is to be maintained.

Although so much development in technologies for detection has already occurred, there is still a real need for even better methods that marry high sensitivity, specificity, and ease of use. Thus, the development of portable, user-friendly devices for on-site testing can make a big difference in monitoring biogenic amines of seafood at any point in the supply chain—from harvest to consumer.

8.4. FACTORS AFFECTING BIOGENIC AMINE FORMATION

The formation of biogenic amines in seafood may be affected by various factors, especially the conditions of storage, processing methods, and extent of microbial activity. Understanding these factors is critical to the design of strategies for control to allow for safer and more wholesome seafood products.

The storage conditions have a huge effect on the formation of biogenic amines in seafood. Temperature is one of the most important factors, and it directly affects microbial growth and enzyme activity (Visciano et al., 2012). Biogenic amines are formed mainly by the decarboxylation of amino acids by microbial decarboxylases. Such enzymes are produced by certain bacteria that multiply under favorable conditions. The increase in temperature will also raise the level of bacterial growth and the enzymatic activity that produces higher concentrations of biogenic amines. Refrigeration is hence required, if not primarily to kill the bacteria then to slow down their metabolism and reduce the formation of these compounds. Maintenance of the seafood below 4°C can greatly suppress the growth of spoilage bacteria and the production of biogenic amines. Poor refrigeration or temperature abuse during storage will enhance spoilage or amine formation and hence be hazardous to the consumers' health.

Processing methods can also impact seafood concentrations of biogenic amines. Different kinds of processing, such as drying, smoking, fermentation, and freezing, may either inhibit or stimulate the formation of these compounds. Drying and smoking may reduce water activity, reducing microbial growth and enzymatic activity and, consequently, decreasing the formation of biogenic amines. On the other hand, if not performed under controlled conditions, these processes may favor the growth and development of particular bacteria that produce biogenic amines. Although fermentation has been a traditional method for the preservation and improvement of flavor in seafood, it may actually result in high levels of biogenic amines if care is not exercised in the process of fermentation. Fermentation is microbiologically mediated, breaking down the components of seafood; therefore, the selection of microbial strains becomes very important. The application of starter cultures with low decarboxylase activity could control the formation of biogenic amines during fermentation.

In the formation of biogenic amines, microbial activity cannot be overemphasized. Such bacteria, particularly those of the genera Enterobacter, Klebsiella, Morganella, Proteus, and Pseudomonas, act to produce the enzymes responsible for decarboxylating the amino acids into biogenic amines. These can either be naturally present in seafood or introduced through contamination during handling and processing. The primary microbial load in seafood and the presence of specific bacteria can be responsible for biogenic amine formation. Good hygiene practices involving the cleaning and sanitizing of equipment and facilities can help retain microbial contamination at bay and reduce the production of biogenic amines.

Moreover, it is the pH of seafood that may alter microbial activity or enzymatic action. Most decarboxylase enzymes exhibit optimal activity at slightly acid to neutral pH values. Changes in pH during processing and storage may favor or hinder the activity of enzymes and microbes. For example, the production of organic acids during spoilage lowers the pH, so conditions favor decarboxylase activity and amine formation. Knowing and controlling the pH could be a good way to control the levels of biogenic amines in seafood.

Type of seafood and intrinsic properties: Many fish and shellfish species contain varying levels of amino acids, which contribute to varying predispositions toward amine formation. Histidine levels are usually higher in fish with high free histidine levels, such as mackerel, tuna, and sardines. Muscle structure and fat content may also influence the level of biogenic amines through their effects on microbial growth and enzyme activity.

Another important factor is the duration of storage. Even if it is stored correctly under refrigeration, there will be a gradual accumulation of biogenic amines after a longer period of time. This is because metabolic activity in spoilage bacteria, though at a slow pace, continues with time. Keeping the time between harvest and consumption at a minimum can go a long way in minimizing the risk of high levels of biogenic amines. Proper supply chain management, including efficient procedures for processing and quickly distributing seafood, could be a solution to fresh seafood retaining its quality and safety. More specifically, preservatives and packaging technologies could also contribute much to controlling the formation of biogenic amines. Preservatives like organic acids, sulfites, or natural extracts might further suppress microbial growth and reduce decarboxylase activity. Other factors, like modification of atmosphere packaging (MAP), may prevent microbial growth and extend shelf life, and hence lengthen the period for which the seafood remains fresh. These methods are combined with proper storage and handling practices to effectively control the level of biogenic amines (Figure 8.4).

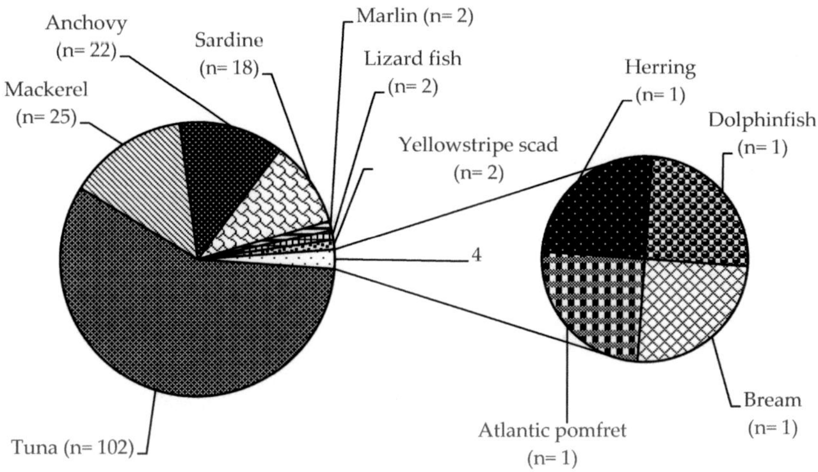

Figure 8.4. *Distribution of histamine in fish species according to the rapid alert system for food and feed from 1 January 2015–31 August 2020.*

Source: Visciano P, Schirone M, Paparella A. An Overview of Histamine and Other Biogenic Amines in Fish and Fish Products. Foods. *2020; 9(12):1795. https://doi.org/10.3390/foods9121795.*

8.5. CONTROL STRATEGIES FOR BIOGENIC AMINES IN SEAFOOD

Biogenic amine formation in seafood must be controlled if food safety and consumer confidence are to prevail. Effective control strategies bring together various preventive measures on handling, storage, preservatives, and packaging technologies. These strategies will help alleviate possible dangers from biogenic amines to the seafood market, allowing it to produce safe products of quality and execute them for consumption by consumers.

Control measures in seafood begin with very strict hygiene and good practices throughout the seafood supply chain (Abuhlega and Ali, 2022). From catch to process and distribution, clean and sanitary conditions are a must. This would include frequent cleaning and sanitization of equipment, surfaces, and storage areas to reduce the microbial contamination of the same. To avoid the ingress and multiplication of bacteria with the capacity to form biogenic amines, training personnel in good hygiene practices is very necessary. Hygiene at the very beginning of processing can be considered one of the most effective ways to reduce the initial microbial load on seafood and, hence, lower the risk of biogenic amine formation.

Temperature control is another important line of strategy in the prevention of biogenic amine formation. This simply means that seafood should always be kept at low temperatures from the time it is caught until consumption. Refrigeration slows down bacterial growth and enzyme activity, hence reducing the production of biogenic amines. Ideally, seafood should be kept below 4°C. Rapid cooling after catching and later transport, processing, and storage along the cold chain are the steps needed to make this practice a reality. Periods of temperature abuse for even short times can lead to rapid growth of bacteria and biogenic amine levels that may increase; therefore, the temperature should always be monitored.

Besides temperature control, the pH of the seafood environment can also become a factor in microbial activity and enzyme function. The control of biogenic amine formation can be ensured by adjustment of the pH to some value that does not favor bacterial growth. This can also be attained by the addition of acidifying agents or other pH-modifying treatments at the time of processing. Monitoring and maintaining the appropriate levels for pH can therefore be an efficient way to inhibit the action of the decarboxylase enzymes responsible for biogenic amine production.

A significant effort put into the control of biogenic amines is in the use of preservatives. Variations of preservatives depress microbial growth and enzymatic activity and, therefore, the formation of biogenic amines. Organic acids, such as acetic, lactic, and citric acids, bring down the pH to an acidic range that does not support bacterial growth. Sulfites are another preservative class whose mode of action includes inhibition of decarboxylase enzymes. However, the use of sulfites is regulated due to potential allergic reactions in sensitive individuals, and their application has to be carefully controlled and labeled.

Natural extracts and antimicrobial essential oils are under investigation for use as preservatives for seafood preservation. Extracted from herbs and spices, these natural preservatives can act to inhibit the growth of bacteria that cause spoilage and reduce the levels of biogenic amines. For example, extracts of thyme, rosemary, and oregano have shown promise in controlling microbial activity regarding seafood. Natural preservatives are according to the consumer's choice of clean-label products and can help in the safety and shelf life extension of seafood.

Other strategies for controlling biogenic amines may be helped by packaging technologies. MAP is one of the widely used techniques, which involves changing the gas composition inside the package, thus inhibiting microbial growth. By substituting oxygen with gases such as carbon dioxide and nitrogen, MAP is able to create an environment where bacteria are proliferating and enzymatic activity is slowed. Similarly, vacuum packaging removes air from the package, making less oxygen available, thereby further inhibiting the growth of aerobic spoilage bacteria. This, when combined with proper refrigeration, extends the shelf life of seafood effectively while maintaining its safety and quality.

Another state-of-the-art type of packaging material could be antimicrobial packaging materials. These are those materials coated or impregnated with antimicrobial agents that inhibit the growth of microorganisms on the surface of seafood. Antimicrobial packaging may give further control against microbial contamination and biogenic amine formation. Research is focused on the development and optimization of these materials so that they can be commercially used to maximize seafood product safety and to prolong shelf life.

This would involve the regular monitoring and testing for levels of biogenic amines in seafood as part of an effective control strategy. Methods for analysis, such as high-performance liquid chromatography and enzyme-linked immunosorbent assay, allow for the measurement of biogenic amine concentrations. Routine testing for control allows the identification of emerging problems and compliance with regulatory limits. A monitoring system such as this allows producers at an early stage to detect and solve problems and thus prevent defective products from reaching consumers.

Biogenic amine control also involves consumer education. Reduction of the risk of biogenic amines can be implemented right from the household level if consumers practice proper handling and storage of seafood. This can be achieved by informing the consumers about the importance of refrigeration, recognizing the signs of spoilage, and avoiding temperature abuse. This will be further strengthened with clear labeling about storage conditions and use-by dates to promote consumer safety and confidence.

8.6. REGULATIONS REGARDING BIOGENIC AMINES IN SEAFOOD

Several health and food safety organizations around the world have proposed regulatory limits for biogenic amines, particularly histamine. The European Union recommended a maximum limit of histamine in fishery products, not more than 100 mg/kg on average in nine samples tested, and none of these samples shall contain more than 200 mg/kg (Ildiko, 2015). The FDA has stipulated an action level of 50 parts per million for histamine in fish in the United States. These limits are based on toxicological studies and risk assessments to prevent histamine fish poisoning, also called scombroid poisoning.

Besides histamine, some legislations also include other biogenic amines like tyramine, putrescine, and cadaverine. Although there is no general limit set for these amines, some countries or regulatory bodies may provide guidelines on their acceptable levels. These guidelines help in the risk assessment efforts and provide a framework for monitoring and controlling biogenic amine levels in seafood.

8.6.1. Enforcement and Compliance Issues

The rules and regulations concerning biogenic amines in seafood are enforced through frequent monitoring, testing, and inspection by food safety authorities. Regulatory agencies frequently check seafood products to determine their content within the set limits. This checking is necessary for detecting and dealing with the potential safety concerns of a product before it reaches the customers.

The main difficulties in the control of the above regulations arise from the variability in the concentration of biogenic amines in different seafood species, handling practices, and storage conditions. All these factors make the establishment of unique limits quite hard to be applied in all situations. Thus, when establishing the limits, control authorities should take into account the above-mentioned factors at the time of inspection.

Compliance with the regulations on biogenic amines by seafood industries is very paramount. It is highly recommended to have both good manufacturing practices, GMPs and Hazard Analysis and Critical Control Points, HACCP systems in place to reduce levels of biogenic amines in producers, processors, and distributors. This would involve the identification of critical points during the production process at which these biogenic amines are likely to form, together with the establishment of control measures to reduce this risk. The proper training of personnel, regular testing, and hygiene are inherent in these control systems. Even with such control measures, there are instances of compliance problems arising for a host of reasons, such as lapses in handling practices, lack of temperature control, or poor monitoring. For small producers and processors, the challenges of installing and maintaining rigid control systems are particularly sharp because they are typically constrained by resources and skill levels. Regulation authorities and industry organizations should continue to support and educate them on how to surmount these challenges.

8.7. EMERGING TECHNOLOGIES FOR BIOGENIC AMINE ANALYSIS

In recent years, techniques for the detection and monitoring of biogenic amines in seafood have greatly improved; certainly, more accurate and sensitive methods are needed. These new technologies make huge contributions toward food safety through the proper control and management of the contents of biogenic amines. These new methodologies of detection open up innovations and future directions in monitoring and improving seafood safety and quality.

Basically, advances in the methods of detection have been targeted mainly at increasing sensitivity, specificity, and speed in the analysis of biogenic amines. In particular, progress in chromatographic techniques was realized through the introduction of ultra-high performance liquid chromatography. UHPLC exhibits several advantages compared to conventional HPLC: higher resolution, faster analysis times, and higher sensitivity. This enables the accurate detection and quantification of biogenic amines at lower concentrations; hence, this would be able to deliver key insight into the early identification and prevention of potential health risks.

One more highly influential development would be the coupling of chromatographic techniques with mass spectrometry. The coupling of either UHPLC or gas chromatography with mass spectrometry results in very powerful methods such as UHPLC-MS or GC-MS. Mass spectrometry provides additional power to the detection of biogenic amines by identifying the molecular structure of these compounds. Their combination allows the analysis of several biogenic amines simultaneously and offers complete profiling concerning their presence in seafood. The very high sensitivity and accuracy offered by these methods make them indispensable tools for compliance and quality control within the seafood industry.

Although chromatographic techniques prevails in the detection of biogenic amines, they have recently been joined by the emerging technology of biosensors. In biosensors, biological recognition elements are used, often in the form of enzymes or antibodies, bound to specific biogenic amines. The binding elements generate a measurable signal that can be an electrical or optical response. Some advantages linked with these biosensors are rapid detection, ease of use, and the potential for on-site testing. This makes them very useful for real-time monitoring during the processing and storage of seafood. Biosensors proved even more useful with the advent of nanotechnology, further increasing their sensitivity and specificity for routine biogenic amine analyses.

Microfluidic devices are another exciting approach to biogenic amine detection; they're often referred to as lab-on-a-chip technologies. These devices integrate a number of laboratory functions onto one small chip, facilitating the miniaturization and automation of analytical processes (Nagasawa, Vora, Agwu, and Tan, 2016). Microfluidic devices offer fast analysis with very minimum sample and reagent requirements. They are capable of running complex biochemical assays within a fraction of the time it takes using conventional methods. The portability and efficiency of microfluidic devices make them very ideal for on-site monitoring, mainly allowing real-time decisions at seafood processing facilities.

Spectroscopic techniques, such as near-infrared and Raman spectroscopy, have also been investigated for biogenic amine analysis. These non-destructive techniques give quick and accurate measurements of biogenic amines with little to no preparation needed for a sample. Light interacts with a sample in the case of NIR and Raman spectroscopy to generate spectra that can be used in analyzing for identification and quantitation of biogenic amines. Such techniques may be applied in continuous monitoring or, equivalently, real-time analysis, hence increasing the chances of detecting biogenic amine formation at an early stage.

Looking forward, there is a huge potential for the integration of artificial intelligence and machine learning with biogenic amine analysis in the future. AI and ML algorithms can be applied to large datasets generated by advanced methods of detection, pointing out the patterns and correlations that might remain elusive through traditional analysis. These will give better accuracy in the detection of biogenic amines and the risks that could possibly arise from historical data. AI and ML can also help to optimize processing conditions and storage practices in a way that minimizes the formation of biogenic amines, thus providing valuable insights into the seafood industry.

Another future direction in the monitoring of biogenic amines is the development of portable and user-friendly field test devices. Portable microfluidic devices and handheld biosensors could place seafood producer, processor, and even consumer on-site testing capability in the hands of users with very little training. This would lead to immediate identification of biogenic amines and thus prompt corrective actions to be taken at the risk of contaminated seafood from entering the market. Another way blockchain technology can further drive a change in the monitoring and traceability of biogenic amines is by using it as a secure, transparent digital ledger to trace seafood products from production to the consumer down the supply chain. This technology captures important data on storage conditions, process steps, and levels of biogenic amine at each stage, thereby making appropriate, correct, and updated information available to each player in the supply chain. Blockchain technology provides better traceability, accountability, and transparency to food, thus increasing safety and consumer confidence.

8.8. CONSUMER PERCEPTION OF BIOGENIC AMINES IN SEAFOOD

The factors that influence consumer perception of biogenic amines regarding seafood include awareness and understanding of the compounds and their health implications. Biogenic amines are a category of naturally occurring compounds that include histamine, tyramine, and putrescine. While they can be associated with some important health risks in high concentrations, there is a very high variation in the level of awareness and related understanding among consumers, which in turn affects purchasing decisions and general confidence in seafood products.

Generally, consumer awareness and understanding of biogenic amines are low. Many consumers are not familiar with the term "biogenic amines" and the possible health risks arising from high levels of these compounds in seafood. Histamine poisoning, also known as scombroid poisoning, is one of the best-known conditions associated with biogenic amines, and even this is not well known to the public. This has partially been because of the limited scope of public education and information on biogenic amines in food safety.

However, with the current trend of increasing awareness about biogenic amines and their associated risk among consumers, they tend to be more careful in handling seafood purchases. The more aware consumers are likely to seek information about handling conditions of seafood during processing and storage. They prefer buying from trusted sources, characterized by good manufacturing practices and very rigorous quality control measures in their processes. Very likely, such consumers would study the labels about storage recommendations and any expiration dates, since they realize that proper handling and storage could play a very important role in minimizing biogenic amine formation.

The attention to biogenic amine awareness can also be viewed in terms of the demand for fresh and well-preserved seafood. A person who understands the risks associated with amines will buy only seafood that obviously looks fresh and whose originators clearly have it properly refrigerated. They might also be interested in buying from retailers who demonstrate proper regard for food safety, such as through advanced preservation techniques or clear communication of handling practices. Additionally, the seafood supply chain faces growing consumer demand for transparency and traceability for food safety reasons. Consumers are developing an interest in knowing where the fish was caught, and how it is handled, processed, and stored. Traceability systems that allow consumers access to detailed information about the journey of seafood from harvest to table will enhance consumer confidence. The transparency allows consumers to make informed decisions and thus reduces their risk of being exposed to high levels of biogenic amines.

The general lack of awareness and understanding of biogenic amines, however, could induce complacency and less cautious purchasing behaviors among the broader consumer base. Since most consumers may not consider the conditions of handling and storage, they often resort to peripheral cues such as price, appearance, and reputation of the brand while making purchase decisions. Doing so may raise their risk toward the consumption of seafood with high levels of biogenic amines, especially if proper handling and storage practices have not been observed.

Education can bridge the knowledge gap for the consumers. Consumers should be sensitized on the existence of biogenic amines and possible health risk factors through effective public health campaigns and education on food safety. Clear, transparent information empowers consumers to make safer buying decisions. This might include some educational materials on the need for proper refrigeration of fish, the signs of spoilage, and possible health implications arising from biogenic amine consumption.

Retailers and manufacturers of food products are also vital in this educative process. Clearly labeling seafood products regarding storage conditions, handling instructions, and expiry will allow consumers to make informed decisions. Retailers can also offer in-store information and staff training programs to clarify consumer doubts regarding seafood safety. This kind of transparency in communication about the steps taken to ensure the safety and quality of seafood will build trust and encourage more cautious purchasing behavior.

8.9. FUTURE PROSPECTS IN BIOGENIC AMINE RESEARCH

With innovative approaches and interdisciplinary collaboration, the future of biogenic amine research in seafood becomes very promising. Several areas of potential innovation and research opportunity have come to the fore accordingly with a growing understanding of the biogenic amines and their health impacts.

One such area of possible innovation is the development of advanced technologies for detection. Improvements to analytical methods, such as ultra-high performance liquid chromatography coupled to mass spectrometry, are increasing in sensitivity and specificity. Other novel diagnostic tools under development include portable biosensors and microfluidic devices that will enable rapid and on-site testing. Those improvements could provide real-time monitoring from harvest to consumer along the seafood supply chain, hence better controlling the levels of biogenic amines and enhancing food safety.

This has multiple possibilities for interdisciplinarity. Interdisciplinary inputs from food scientists, microbiologists, chemists, and technologists can improve preservation methods and storage technologies. For example, studies on microbial ecology coupled with advances in food technology may lead to the invention of new preservation methods that would hinder biogenic amine formation. Therefore, some studies on the genetic and enzymatic pathways of biogenic amine-producing bacteria could discover new strategies to scotch their activity.

Another critical area of research is in the domain of the effects of biogenic amines on human health. Interdisciplinary research among toxicologists, nutritionists, and medical professionals into the mechanisms involved in the action of biogenic amines on health shall help risk assessments and regulatory standards; only then, with an understanding of individual susceptibilities and long-term effects of low-level exposure, would dietary guidelines and public health recommendations be refined.

Furthermore, consumer education and risk communication represent indispensable research areas. There, behavioral scientists and public health experts can devise strategies for effectively communicating the risks associated with biogenic amines and promoting safe handling practices in seafood.

Chapter 9

Seafood Toxins

Contents

INTRODUCTION

Seafood is an integral part of diets all over the world and is valued for its high nutritional value as well as variety in food. At the same time, seafood creates some of the biggest food safety issues due to the presence of a number of toxins. The source of the toxins could be natural, like marine biotoxins coming from algae, or chemicals, resulting from human activities leading to contamination of the marine environment. Understanding seafood toxins is very critical in ensuring that the consumption of seafood is safe for human health protection. Marine biotoxins produced by particular types of algae and microorganisms settle in tissues of filter-feeding shellfish and other marine organisms, which can later cause severe sicknesses in human beings, from gastrointestinal to neurological disorders. Chemical contaminants of seafood, associated with bioaccumulation and biomagnification, include heavy metals like mercury and lead, industrial pollutants like polychlorinated biphenyls, and pesticides. It has been revealed that the potential health risks associated with such chemical contaminants are long-term, particularly developmental and reproductive toxicity, immunotoxicity, and carcinogenicity.

Seafood toxins detection and management is very complex and calls for robust scientific and regulatory frameworks. Advanced analytical techniques are required in the identification and quantification of these toxins in seafood. Monitoring and surveillance programs are capable of detecting the presence of toxins in advance and preventing contamination. Regulatory agencies charged with setting standards and guidelines of safety, ensuring compliance, and assessment of risks will play a cardinal role in protecting consumers. This also entails being aware of public education regarding seafood toxins as part of the food safety system. Consumers should be informed of the risks associated with seafood consumption and that seafood should be acquired from safe and reputable suppliers. Media and communication strategies would increase public awareness in support of good eating practices.

Research and innovation are the first to meet the real challenges seafood toxins pose. The current trends in research on seafood toxins are oriented to the discovery of new toxins, the development of more sensitive and rapid detection methods, and the understanding of ecological and environmental factors affecting toxin production. This requires interdisciplinarity among scientists, regulators, and industry stakeholders in furthering knowledge and developing efficient approaches to mitigate the risks associated with seafood toxins.

9.1. TYPES OF SEAFOOD TOXINS

Seafood toxins can be broadly categorized into biotoxins and chemical contaminants. Both types of toxins pose significant health risks to humans and have diverse sources and mechanisms of action.

9.1.1. Biotoxins

Biotoxins are naturally occurring toxic compounds produced by some species of marine algae and microorganisms. These toxins accumulate in marine organisms, mainly in filter-feeding shellfish, by feeding upon the algae which produce the toxins. Biotoxins can lead to serious and sometimes fatal intoxications in humans, resulting from the consumption of marine organisms that have consumed the toxic algae (Figure 9.1).

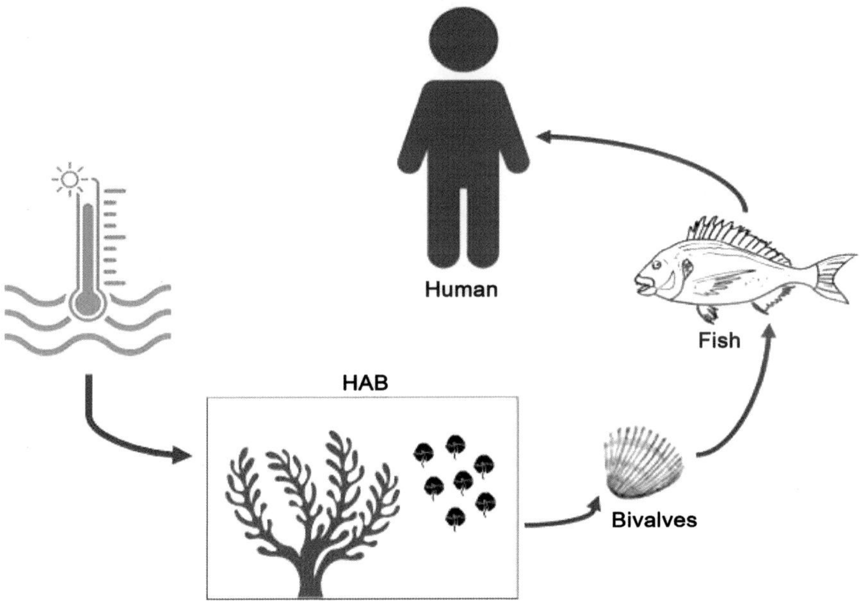

Figure 9.1. *Marine biotoxins.*

Source: Barbosa V, Santos M, Anacleto P, Maulvault AL, Pousão-Ferreira P, Costa PR, Marques A. Paralytic Shellfish Toxins and Ocean Warming: Bioaccumulation and Ecotoxicological Responses in Juvenile Gilthead Seabream (Sparus aurata). Toxins. 2019; 11(7):408. https://doi.org/10.3390/toxins11070408).

The general major classes of biotoxins are:

- **Paralytic Shellfish Toxins (PSTs):** These are mainly produced by dinoflagellates, especially from the genus Alexandrium. Among PSTs, saxitoxins and their analogs block sodium channels in nerve cells and cause paralysis. There are symptoms of PST intoxication called paralytic shellfish poisoning (PSP), which is usually characterized by tingling, numbness, dizziness, and possible respiratory failure or even death in extreme cases (Etheridge, 2010). Poisoned bivalves and mussels, which are from contaminated shellfish, are the major sources of this group of toxins.

- **Amnesic Shellfish Toxins (ASTs):** Domoic acid is one of the examples of this group, and it is produced by Pseudo-nitzschia. ASP is a problem seen with people who have ingested domoic acid. Symptoms of ASP can include gastrointestinal upset and neurological effects such as memory loss, confusion, and seizures. In severe cases, it may cause coma or death. Shellfish and some fish that feed on toxic algae can concentrate ASTs.
- **Diarrhetic Shellfish Toxins (DSTs):** DSTs are produced by various dinoflagellates, among them Dinophysis and Prorocentrum, and are primarily made up of okadaic acid and related derivatives. These toxins inhibit protein phosphatase and cause the illness known as diarrhetic shellfish poisoning (DSP), a consequence of consumption with diarrhea, nausea, vomiting, and abdominal pain. Although most cases of DSP are not lethal, they can have bothersome and disturbing health effects. Shellfish are the vectors of DSTs.
- **Neurotoxic Shellfish Toxins (NSTs):** These toxins, produced by dinoflagellates of the genus Karenia, include neurotoxins called brevetoxins. NSP is the abbreviation for neurotoxic shellfish poisoning. Symptoms include gastrointestinal effects and neurological effects that include nausea, vomiting, diarrhea, and paresthesia. NSTs may cause respiratory difficulties if aerosolized brevetoxins are inhaled. The most common source of NSTs is shellfish that concentrate brevetoxins.
- **Ciguatoxins:** These are produced by dinoflagellates from species of Gambierdiscus and get concentrated in coral reef fish such as barracuda, grouper and snapper. The characteristic feature of ciguatoxins is ciguatera fish poisoning (CFP), which exhibits symptoms of the digestive, neurological and cardiovascular systems. Symptoms of CFP include nausea, vomiting, diarrhea, tingling, temperature reversal (hot feels cold and vice versa) and in some cases, cardiac and/or neurologic dysfunction. Unlike other shellfish toxins, ciguatoxins are most commonly associated with fish.

9.1.2. Chemical contaminants

Chemical contaminants in seafood mainly stem from industrial pollution, agricultural runoff, and lack of proper waste disposal. These residues have been noted to inundate marine environments and filtrate through the food chain and into human beings for whom they present serious health dangers. Major types of chemical contaminants include:

- **Heavy Metals:** Such metals as mercury, lead, cadmium, and arsenic are common pollutants that accumulate in seafood. It is the neurotoxic effect that represents one of the reasons mercury is particularly problematic, mainly in its most harmful forms, like methylmercury, which affects the developing nervous system. Chronic mercury exposure can result in neurological damage, developmental delays in children, and cardiovascular disease. Lead exposure may cause neurological and developmental problems, whereas cadmium and arsenic are usually linked to the damage of renal systems and cancers. Fish and shellfish can accumulate metals, the higher-order predatory fish like tuna and swordfish typically have higher concentrations of mercury.
- **Polychlorinated Biphenyls (PCBs):** These are industrial chemicals used in large quantities in electrical equipment, paints, and other products. Their use has been recently prohibited in most countries. They have properties of not breaking down easily and, as such, can accumulate in fish and marine mammals' fatty tissues. PCBs are known to be cancer-causing and cause immune system, reproductive, and endocrine system disruptions. They are present in seafood, mainly fatty fish, and thus pose a threat in the food chain to humans.
- **Pesticides:** Many of the agricultural runoffs contain pesticides, which can eventually deposit in marine environments and also accumulate in seafood. Organochlorine pesticides, like DDT, are persistent in nature and have severe health effects for the occurrence of cancers, reproductive impairment, and endocrine disruption. Organophosphate pesticides are known to cause acute toxicity ranging from nausea and dizziness to muscular anesthetizing to the extent of seizures and respiratory failure.
- **Dioxins and Furans:** Dioxins and furans are created in many industrial activities, including waste incineration and chemical production. Because of their persistence, dioxins and furans easily bioaccumulate. They are confirmed carcinogens that can cause immune system diminution, reproductive and developmental disturbances, and endocrine function disruption. Major dietary

sources of dioxins and furans include fish and shellfish, primarily from contaminated zones.

- **Polycyclic Aromatic Hydrocarbons:** These are by-products of incomplete combustion of organic matter, introduced into the marine environment from oil spills, industrial discharges, and runoff. Some PAHs are known to be carcinogenic and could be expected to cause respiratory problems, skin irritation, and reproductive difficulties. PAHs deposit in seafood, predominantly in shellfish, from contaminated sediments and water.

9.2. SOURCES OF SEAFOOD TOXINS

Seafood toxins have diverse sources, both environmental and biological, with most of these credited to or influenced by human activities. The majority of the toxins are produced by microorganisms like algae and bacteria in the sea. These organisms secrete naturally occurring biotoxins as by-products of their metabolism.

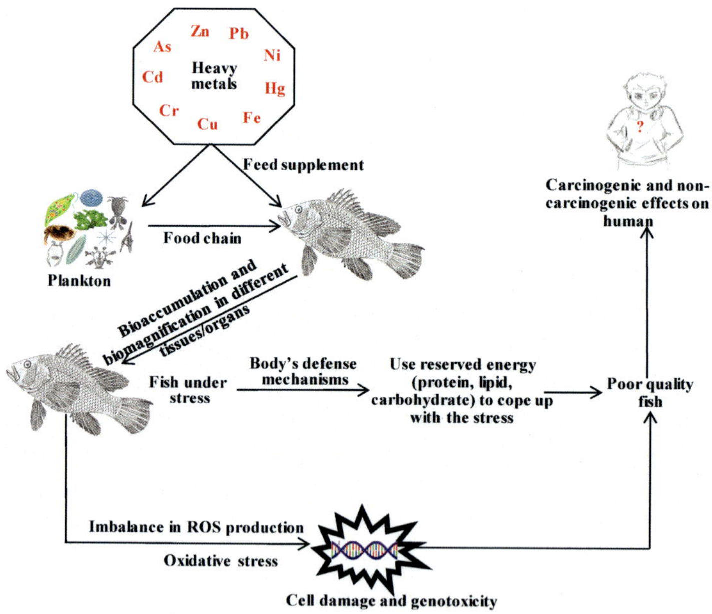

Figure 9.2. *Human health risks associated with consumption of fish contaminated with trace elements from intensive mining activities.*

Source: Jamil Emon F, Rohani MF, Sumaiya N, Tuj Jannat MF, Akter Y, Shahjahan M, Abdul Kari Z, Tahiluddin AB, Goh KW. Bioaccumulation and Bioremediation of Heavy Metals in Fishes—A Review. Toxics. *2023; 11(6):510. https://doi.org/10.3390/toxics11060510.*

The best example is dinoflagellates and diatoms that produce harmful algal blooms: the by-products of which are biotoxins discharged into the marine environment. Filter-feeding shellfish genera, i.e., clams, mussels, and oysters take up these poisonous algae and accumulate the toxins in their bodies. The phenomenon of accumulation of a toxin, no matter how minute in dosage, according to science, poses an astonishing peril to human health upon the consumption of the contaminated shellfish (Figure 9.2).

There is a complex relationship between marine biotoxins and paralytic shellfish poisoning (PSP), amnesic shellfish poisoning (ASP), and diarrhetic shellfish poisoning (DSP). In fact, these are the biotoxins that come forth from these natural processes. They can have really bad health outcomes such as neurologic symptoms and gastrointestinal symptoms. Now, there are also several environmental factors that play a role in the onset and severity of a harmful algal bloom, which include water temperature, salinity, and nutrient availability. Climate change and natural climatic events such as El Niño can worsen this frequency and distribution, increasing the risk of exposing themselves to toxins in seafood.

Besides these naturally occurring biotoxins, seafood can also be contaminated with chemical pollutants from anthropogenic origin. Industrialization, urbanization, and agricultural practices significantly enhance the incidence of these pollutants in marine environments (Marques, Maulvault, and Nunes, 2019). Mercury, lead, cadmium, and arsenic are some of the common heavy metal pollutants discharged into the sea from various industries, mining activities, and agricultural lands. Because they are not decomposed by enzymes, such metals accumulate in the tissue of marine organisms, offering a potential threat to the tissue consumed.

Mercury, particularly methylmercury, is such a bioaccumulating contaminant to the marine food chain. Therefore, owing to their higher position in the food web, increased levels of mercury are found in larger predatory fish than in smaller ones, such as tuna, swordfish, and shark. Exposures to high mercury intakes may result in neurotoxicity or delays in development, especially in fetuses and young children.

Other chemical contaminants arising from human activities include polychlorinated biphenyls, dioxins, and furans. These are by-products of industrial processes, including the manufacture of chemicals and burning of wastes, which are then released into the environment where they take a long time to break down. Before being banned in most countries, PCBs were used in many electrical equipment, paint, and plastics. However, they are environmentally persistent and bio-accumulate in the fatty tissues of fish and marine mammals. Dioxins and furans are both by-products of combustion processes, tend to accumulate in the food chain as well, and are known to have effects on cancers, reproductive problems, and immune suppression.

Aggressions of agricultural activities introduce contamination into the seafood through the runoff, mainly pesticides, and fertilizers, which are absorbed by marine organisms and may impart several health effects. The DDT and organophosphate group of pesticides may accumulate and cause endocrine disruption and neurotoxicity. The fertilizers further add to excess nutrients that may enhance eutrophication, thus supporting the growth of harmful algal blooms and increasing the risk of exposure to biotoxins.

Human activities not only introduce marine contaminants into the sea but also worsen the conditions that lead to the growth of natural toxins. Coastal development, overfishing, habitat destruction, and other human activities can lead to the disruption of marine ecosystems, causing a shift in the balance of microbial communities that may favor toxin-producing organisms. Global climate change is producing an alteration in sea surface temperature, currents, and sea levels, giving rise to conditions that facilitate the growth and spread of harmful algae.

Other anthropogenic activities are marine shipping and oil spills, which pose a contamination of seafood. Oil spill introduces PAHs into the marine ecosystem, absorbed by marine wildlife organisms. PAHs are carcinogens that can cause respiratory and reproductive damage. Cleanup and remediation during an oil spill are complex and cause severe chronic impacts on the environment.

9.3. HEALTH EFFECTS OF SEAFOOD TOXINS

The health effects of seafood toxins are very diverse and depend on the type, concentration of the toxin, and the individual's susceptibility. Acute seafood toxin poisoning presents with sudden, serious, immediate symptoms after ingestion. One of the best-known examples is histamine poisoning, also known as scombroid poisoning. This happens because fish may be stored poorly, which allows for the growth of bacteria that turn histidine in fish into histamine. The symptoms of poisoning by histamine include flushing, rash, headache, dizziness, nausea, vomiting, and diarrhea. Severe symptoms include respiratory distress and palpitations. Treatment typically includes antihistamines for symptomatic management. However, patients with severe cases will require hospitalization and supportive care.

Another acute condition is PSP, which is an acute neurotoxic illness due to the intake of shellfish that have been contaminated with saxitoxins produced by particular algal species. Typical symptoms and signs include tingling and numbness around the lips and at the extremities, dizziness, headache, and muscular paralysis, which in severe cases may lead to respiratory failure. There is no specific antidote for PSP, so treatment is supportive and includes respiratory support if necessary. Hence, early seeking of medical attention is very important to ensure the prevention of fatal outcomes (Figure 9.3).

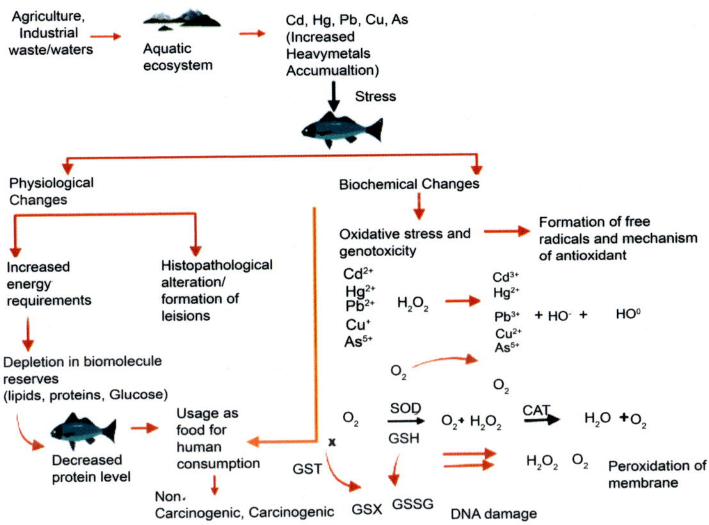

Figure 9.3. *Ocean contamination with trace metals and its impact on human health through seafood consumption.*

Source: Naz S, Chatha AMM, Téllez-Isaías G, Ullah S, Ullah Q, Khan MZ, Shah MK, Abbas G, Kiran A, Mushtaq R, et al. A Comprehensive Review on Metallic Trace Elements Toxicity in Fishes and Potential Remedial Measures. Water. 2023; 15(16):3017. https://doi.org/10.3390/w15163017.

Domoic acid is the production of the diatoms of the genus Pseudo-nitzschia. In human food, this poison causes amnesic shellfish poisoning, which includes gastrointestinal and neurological symptoms (Kumar, Kumar, and Nair, 2009). At its very beginning, the symptoms are nausea, emesis, diarrhea, and abdominal cramps. The condition may further develop into confusion, memory loss, seizures, and even, in some serious cases, coma or death. The treatment for ASP is not specific; it needs management by supportive care and monitoring of neurological complications.

Diarrhetic shellfish poisoning is a toxic condition caused by the consumption of shellfish that are contaminated with okadaic acid and all its congeners. In this case, symptoms are related to the gastrointestinal tract: diarrhea, nausea, vomiting, and abdominal pains. The symptoms usually go away in several days without any specific treatment. However, rehydration and replacement of lost electrolytes may sometimes be necessary in extreme cases.

The very long-term health effects may result from chronic exposure to seafood toxins and are generally caused by the accumulation of toxins in the body. Among the seafood toxins, mercury, mainly in its methylated form, is considered to present the greatest neurotoxic risk. Chronic mercury exposure may cause neurological and developmental problems such as cognitive deficits, motor dysfunction, and sensory impairments. It is most susceptible in pregnant women and small children. Chronic mercury exposure is managed by dietary adjustments to limit the intake of fish that contains a high level of mercury together with regular monitoring of its level in the body.

Polychlorinated biphenyls and dioxins are persistent environmental pollutants bioaccumulating in the fatty tissues of fish. Chronic exposure to these compounds has been linked to immune suppression, reproductive and developmental problems, endocrine disruption, and increased cancer risk. As these toxins can remain in the body for years, long-term effects are of concern. Mitigation measures include selecting seafood with lower fat contents and avoiding fish from contaminated waters.

Ciguatera fish poisoning is caused by ciguatoxins from marine dinoflagellates, then accumulated in reef fish. Ciguatoxins ingested may produce acute and chronic symptoms. The acute syndrome includes gastrointestinal, neurological (reversal of temperature sensation: hot feels cold and vice versa), muscular, and cardiovascular symptoms. Chronic effects may last for weeks or months and include asthenia, depression, and neurological disorders. There is no specific treatment for CFP; management focuses on symptom relief and supportive care.

The successful management of medical impacts caused by seafood toxins is founded on an early diagnosis and early intervention. Providers should be knowledgeable about the symptoms and probable origins of seafood toxins so appropriate care can be provided. This also includes very important public health initiatives aimed at educating consumers about risks associated with seafood toxins and promoting safe consumption practices.

9.4. DETECTION OF SEAFOOD TOXINS

High-performance liquid Chromatography technique (HPLC) is among the most used in detecting seafood toxins. HPLC separation allows the separation, identification, and quantitation of multiple toxin compounds in a single analysis (Turner, 2014). The methodology is very effective at detecting biotoxins present in seafood. Its high sensitivity and specificity make HPLC one of the methods of choice for regulatory testing and quality control within the seafood industry.

However, HPLC is associated with sophisticated equipment and well-trained personnel, and extensive sample preparation is considered time-consuming and expensive.

When combined with chromatography techniques such as HPLC or gas chromatography, mass spectrometry improves the detection capabilities of seafood toxins. Mass spectrometry provides detailed molecular information that can enable the exact identification of the structure of toxins and their metabolites. Techniques such as LC-MS and GC-MS have extremely high accuracy and sensitivity, so they are very useful in confirming the presence of certain toxins. However, these techniques also have complex instrumentation and require great expertise to use and interpret results.

Another methodology very popularly used for the detection of seafood toxins is ELISA. ELISA is based on antibodies for toxin detection and generates a measurable signal proportional to the concentration of target compounds. This technique is relatively easy, quick, inexpensive, and thus quite applicable to routine screening. It is of particular use for the detection of domoic acid and saxitoxins in shellfish. However, the technique is only as specific as the antibodies used, and it does not provide the same degree of detailed information given by chromatographic methods.

Biosensors are an emerging technology in seafood toxin detection. They consist of analytical devices in which biological recognition elements, enzymes or antibodies, recognize toxins and result in an electronic signal. One of the main advantages of biosensors is their quick, in-process testing, which often comes in handy during real-time monitoring and decision-making. Advances in nanotechnology further improved the sensitivity and specificity of biosensors. Therefore, their application in the seafood industry seems highly promising. However, their development and wide use are still developing and more validation is needed to check their reliability.

One of the larger issues associated with seafood toxin detection is the diversity and complexity of toxins themselves. There can be a variety of biotoxins, apart from chemical contaminants, in seafood that co-occur, all having varying characteristics and health impacts. This then needs an analytical technique that has great versatility for detecting many toxins simultaneously. The development of such multi-analyte methods is technically challenging and usually requires a trade-off between sensitivity, specificity, and throughput.

Another issue is the matrix effect, in which complicated seafood composition can impact toxin detection and quantification. There are various proteins, fats, and other compounds in seafood that might have an impact on the performance of an analytical method. Matrix effects are usually minimized through time-consuming sample preparation and purification steps, increasing the time and cost of analysis. Researchers are constantly looking at ways to better sample preparation techniques so that these effects can be minimized and accuracy maximized.

It is also of interest for efficient monitoring and risk management to be able to apply detection methods rapidly and reliably. While traditional laboratory-based techniques may be accurate, results from them may not arrive quickly enough to prevent contaminated seafood from reaching consumers. Developing portable and user-friendly devices that enable on-site testing can help solve such an issue by providing immediate detection and response. However, one of the biggest challenges is to ensure the same accuracy and reliability with these devices as with the laboratory-based methods.

9.5. PREVENTION OF SEAFOOD TOXIN CONTAMINATION

Prevention of seafood toxins majorly involves the management of risks by the implementation of effective risk management strategies, monitoring, and surveillance mechanisms that identify the potential sources of contamination and manage such risks to see that the seafood reaching the consumers is safe for consumption.

In preventing seafood intoxication due to toxins, we ought to consider the entrance pathways into the seafood supply chain comprehensively. Biotoxins from marine organisms and chemical contaminants of anthropogenic origin are considered the main sources of intoxication, so they should be used as direct measures for minimizing the chances of contamination.

One of the major strategies for the control of biotoxin contamination risk involves the establishment of harmful algal bloom (HAB) monitoring programs. Marine biotoxins (saxitoxins, domoic acid, and brevetoxins) occur mainly as a result of bio-concentration by shellfish and other marine products in areas affected by harmful algal blooms. Monitoring programs for ASPs include regular sampling of water and shellfish taken at these known HAB hot spots. Early detection of HABs ensures that timely management actions are taken, such as the closure of affected harvesting areas, public notifications, and warnings and prevention of seafood products possibly contaminated from entering the market.

Besides HAB monitoring, great importance lies in managing the environmental factors that promote the growth of toxin-producing algae. In this regard, nutrient management through agricultural field runoff, sewage discharge, and other sources of pollution heavily causes algal bloom and eutrophication. Adoption of best management practices (BMPs) for agriculture, increased wastewater treatment, and the encouragement of sustainable coastal development can decrease the number and intensity of HAB events, thereby reducing biotoxin contamination of seafood.

Risks from chemical contaminants such as heavy metals, PCBs, and pesticides are controlled and reduced by managing the sources of related pollution. This usually calls for strict regulatory regimes that channel the release of toxic chemicals into marine environments in very minimal quantities. Circumstances such as these are the ones in which enforcement measures should place monitoring systems on industrial discharges and agricultural runoff to achieve compliance and to be in a constant course of reducing the contamination of the marine ecosystems. The identification and clean-up of polluted sites, such as in former industrial areas and polluted river estuaries, further prevent the introduction of chemical poisons into the food supply chain.

Effective monitoring and surveillance programs are essential for protecting seafood from contamination by toxins. Such programs should involve the collection, analysis, and interpretation of data aimed at seafood safety at different intervals. Monitoring programs that are well established will follow the variation of presence of the toxic substances in the marine environment and seafood commodities. That will consequently be the leading source of essential information for risk assessment and risk management.

For biotoxins, the surveillance program generally includes *in situ* monitoring of the algae that produce toxins aside from laboratory analysis of seafood samples. Determination and quantification of the level of toxins are achieved through advanced techniques in analytical chemistry, mainly the HPLC and MS methods, which make the relevant data very accurate and reliable, so the authorities will be able to make sound decisions about the safety of seafood. Regular testing is needed on seafood from areas, particularly those that are prone to HABs, to provide early biotoxin detection and prevent the contamination of products reaching consumers.

Chemical contaminants are monitored through programs entailing the collection of water, sediments, and marine organisms samples to establish the levels of pollutants. Such programs trace contaminant trends, evaluate the effectiveness of controls as well as guide the development of measures. Surveillance information also supports risk assessments used in setting safe levels of consumption of a variety of seafood to guarantee that the regulatory limits are founded on scientific soundness.

Public health authorities have an important role in monitoring and surveillance, not only for contaminants but also for seafood toxins. Their tasks involve the running and coordinating of monitoring programs, information analysis, and the relay of such information to all stakeholders, including seafood producers, regulators, and the public. They must communicate to build and keep consumer trust and transparency and be in a position to provide information timely and accurately during incidents of contamination.

Besides regulation monitoring, the seafood industry itself needs to take initiatives in the area of toxin contamination prevention. The HACCP systems in the supply chain could provide a system for pinpointing potential hazards, setting critical points of control, and setting measures for preventing or controlling the risks. On the other hand, regular training and education of seafood handlers and processors in the best practices of hygiene, storage, and handling will minimize the risks of contamination.

Consumer education is one other way through which seafood toxin contamination can be prevented. Educating consumers about the dangers associated with seafood toxins, through safe food-handling practices and sourcing seafood from reputable suppliers, hence reduces cases of illness from toxins. Awareness can further be enhanced through public health campaigns, websites, and labeling initiatives.

9.6. REGULATIONS REGARDING SEAFOOD TOXINS

The regulations on seafood toxins have been established to ensure a reduction to acceptable levels. The regulations for acceptable levels of all different toxins in the seafood are issued by both international and national bodies.

The guidelines for the international levels of seafood toxins are set by the agencies or bodies like the Codex Alimentarius Commission, which partners with WHO (World Health Organization) and FAO (Food and Agriculture Organization) of the United Nations (Toyofuku, 2006). The purpose of the Codex is to lay down international food standards, guidelines, and codes of practice to be adopted by the Member countries; it ensures food safety and provides common ground for facilitating international food trade.

To provide an example, some marine biotoxins have been set maximum levels for, e.g., PSP toxins, ASP toxins, and DSP toxins, by the Codex, in relation to shellfish. Based on scientific risk assessment, the levels are created to protect the consumers from harmful effects sourced from the toxins. The Codex also provides for guidelines on monitoring and control of risks associated with HABs, which are the principal source of these toxins.

At the country level, the national regulations are promulgated based on international guidelines and national situations. In the USA, the FDA sets up regulatory limits for some seafood toxins including histamine in fish and biotoxins in mollusk shellfish. The FDA merely sets an action level for histamine in fish, for instance, at 50 parts per million. Fish having this level in it are adjudged adulterated and not good for consumption. The FDA also enforces another requirement that shellfish be harvested from waters that are monitored for biotoxins. It also oversees the state-run programs providing for shellfish sanitation to ensure compliance.

In the European Union, the European Food Safety Authority takes the responsibility of providing scientific advice on food safety, including risks posed by seafood toxins. It is also important to note that the maximum limits, as in the Codex recommendations, have been set for shellfish products in the European Union, in terms of marine biotoxins. The EU also has regulations on monitoring and/or controlling contaminants like heavy metals, dioxins, and PCBs in seafood. These regulations have to be enforced by the member states, where compliance is monitored at very short intervals between inspection visits, supported by testing.

Compliance with the above regulations is implemented through industrial self-regulation supplemented by different levels of government control. It is the responsibility of producers, processors, and distributors to ensure that the seafood products reaching consumers are safe for consumption. It includes compliance with good manufacturing practices (GMPs), the application of Hazard Analysis and Critical Control Points (HACCP) systems, and testing for toxins at regular intervals. Specifically, HACCP comprises a preventative system identifying crucial spots in the production line where hazards can be curbed or eliminated to secure the safety of the final product.

Government agencies are critical in regulatory enforcement and compliance for any country. They perform inspections, monitor seafood products, and test for toxins. If any non-compliance is found, actions can be taken—warnings, fines, pulling the lot, and in grave cases, judicial action to protect the health of the public.

The public health agencies also run surveillance programs geared at keeping track of seafood toxins' incidence and the magnitude of the impacts on human health. Such programs will, therefore, provide data on the levels of seafood toxins and foodborne diseases. The collected data is used to fine-tune the regulations and enhance strategies concerning risk management. The public health agencies also work in partnership with other stakeholders in the food safety chain, such as industrial groups, research institutions, and international bodies, to enhance food safety practices that will curb the emerging risks.

Education and training are important components of compliance and enforcement. Through education and training programs, both governments and industry organizations offer all classes of handlers and processors of seafood as well as inspectors an understanding of safety regulations of these products and compliance with these rules. These programs address such subjects as good hygiene practices, safe handling practices, and the use of analytical methods for the detection of toxins. Such programs increase knowledge and skills so as to preclude contamination and guarantee that the seafood products are considered safe.

Other contributors to seafood safety are the awareness and education of consumers. Public health campaigns and information resources make consumers aware of the risks of seafood toxins and ways of reducing them. For example, buying seafood from reputable sources, storing, and handling it properly and avoiding advice regarding seafood safety. Informed consumers can make better choices and contribute to overall food safety.

9.7. EMERGING TRENDS IN SEAFOOD TOXIN RESEARCH

One of the most invested areas in seafood toxin research is the discovery of new toxins. With an enhanced understanding of marine ecosystems, scientists are now discovering previously unknown toxins produced by algae and other marine organisms. These are often derived from research on HABs and other anomalous marine events. For example, new types of toxins from dinoflagellates and diatoms have been described, which, when accumulated in seafood, may present a risk to human health. Knowing these new toxins itinerates in a better way the formulation of monitoring and management strategies to prevent their entry into the food supply by scientists.

Not only are new toxins being isolated, but scientists are also greatly improving their understanding of how these toxins impact human health, including investigations into mechanisms through which the toxins interact with human cells and organs. It is this type of research that can help to establish truly effective treatments for toxin exposures and provide more accurate risk assessments. With a better understanding of how toxins cause harm, it is possible to create guidelines regarding safe consumption levels and make public health recommendations much more efficient.

Another growth opportunity in this area is the technological improvement being made in seafood toxin detection methods. The traditional methods of detection—HPLC and ELISA have started to be supplemented by more sensitive, new methods. For example, mass spectrometry can be very precise and is also able to detect several toxins at once. This technology is also becoming more available and is used to detect even the smallest traces of toxins in seafood.

Biosensors are the latest development in toxin detection. These devices use biological components like enzymes or antibodies and can effectively and efficiently detect the toxins. These can also be portable biosensors that provide results in real-time—something very useful during field measurements or seafood processing. This real-time feedback prevents contaminated products from reaching consumers.

Another very promising area in the field of nanotechnology is that related to toxin detection. Nanomaterials can enhance the sensitivity and specificity of the applied methods of detection. For example, nanoparticles can be used to amplify signals in biosensors and thus allow for the detection of very low levels of toxins. These kinds of developments could, in the end, lead to far more efficient, cost-effective test methods that are available for everybody.

Genomic and molecular methods have also been applied in seafood toxin research. Scientists work on the genes of organisms that produce toxins and develop molecular assays to detect specific toxins quickly and efficiently. Among these molecular approaches are PCR, polymerase chain reaction, and NGS, next-generation sequencing. They are ways through which from water samples, researchers can identify the presence of toxin-producing algae, serving as a watch-out system for possible seafood contamination.

Big data and AI prove to be very instrumental tools in the public health domain when it comes to monitoring seafood toxin outbreaks and making predictions. AI applied to large datasets emanating from environmental monitoring programs, health records, and seafood testing can aid in detecting patterns and, hence, in predicting future toxin events. This forward-looking approach provides a better strategy for preparedness and responses to reduce the risks of wide-area contamination.

Increased collaboration among international and national agencies is raising the efficacy of seafood toxin research. Cross-border sharing of data and the use of shared facilities can achieve a broader view of the global trends of these toxins. Moreover, international cooperation enables the establishment of standardized testing methods and regulatory frameworks, which ensure consistent standards of seafood safety worldwide.

Research findings also raise the level of awareness among the general public. When researchers discover more about seafood toxins and how to better detect them, this knowledge is shared through education campaigns and resources that extend to the public. The public becomes more knowledgeable about the hazards associated with seafood toxins and ways of mitigating these dangers. One of the most significant benefits of heightened awareness is that it furthers the consumption of safe seafood.

9.8. PUBLIC AWARENESS OF SEAFOOD TOXINS

Educational efforts are one way to create awareness about seafood toxins. Government agencies, health organizations, and community groups can conduct such programs. They attempt to sensitize the nature of toxins in seafood, their effects on health, and ways through which safe consumption can be achieved. For example, food safety education can be incorporated into the learning curriculum; students, at an early age, can be trained on handling and storing seafood properly. Similarly, community workshops and seminars will enable the effective dissemination of information to adults on measures they could take to detect and prevent seafood from contamination.

Another good teaching aid would be public health campaigns. Such publicity should, therefore, use various media, right from simple brochures and posters to websites and many social media platforms. Some of the key messages may consist of very important advisories on the purchase of seafood only from credible sources, their identification of spoilage signs, and safe cooking methods. By communicating such messages far and wide, public health campaigns could ensure a greater understanding of the risks involved and how people can protect themselves.

The media plays a very significant role in communicating information about seafood toxins. Current seafood safety issues can be reported on through news, television programs, and online news. This means that outbreaks of seafood poisoning or updates about new research findings can be presented at appropriate times to the public. By giving timely and appropriate information, the media keeps the public abreast of the potential risks and how one may mitigate them.

Social media has emerged as an increasingly important tool in seafood toxin awareness. Health organizations and experts use such channels as Facebook, Twitter, and Instagram to post information and engage directly with the public regarding marine toxins. Social media can be used to provide tips to safely consume seafood, alert people to new risks, and respond to community questions. Such an interactive approach would help to ensure that all necessary information is provided in formats that are both accessible and easily understandable to people.

Influential personalities and celebrity endorsers may also contribute much towards raising public awareness. When popular figures come out to share their views on seafood safety, they engage large audiences by making the issue more relatable. Their endorsement can influence people to start taking food safety seriously and follow better practices in their lives.

Government agencies play a key role in coordinating efforts on education and dissemination of accurate information to the public. Indeed, these government agencies, like the FDA in the United States or the EFSA in Europe, frequently publish guidelines, reports, and alerts related to seafood safety. They also collaborate with other groups in developing and distributing educational materials.

The other major area of public awareness is that of consumer labeling. Clearly labeling seafood products can help consumers learn about potential hazards and safe-handling methods. Such labels can show the origin of the seafood, potential allergens, and how they should be stored and prepared. All this helps in making informed decisions and reduces the risks of consuming contaminated seafood.

9.9. FUTURE DIRECTIONS IN SEAFOOD TOXIN ANALYSIS

One of the most interesting fields of research is the development of high-throughput screening methods, which include next-generation sequencing and state-of-the-art mass spectrometry. Such techniques allow the fast and exact identification of several toxins simultaneously, increasing the efficiency and reliability of seafood safety testing. For instance, LC-MS and GC-MS are tremendously more sensitive and selective, and hence they allow for the accurate determination of minute residues of toxins.

Another example of an innovation is input from biosensors and nanotechnology in toxin detection. The use of biological elements, such as enzymes or antibodies, to detect toxins gives fast and reliable results. Nanomaterials enhance sensitivity and specificity, making the sensors fit for real-time monitoring all the way from seafood processing plants down to the market stalls. This real-time ability in such is very vital for instant decision-making and preventing contaminated seafood from reaching consumers.

Equally important is interdisciplinary collaboration that advances seafood toxin analysis. Interdisciplinary approaches to research, coupling marine biology, chemistry, toxicology, and bioinformatics, can improve knowledge of how toxins impact seafood and human health. This could be integrated into the development of more practical techniques for the identification of such toxins and risk modeling. For instance, a marine biologist would discover new organisms that could produce toxins, while a chemist develops the analytical methods to detect these toxins, and a bio-informatician for the analysis of data to predict potential outbreaks. In addition, seafood safety testing can be integrated with environmental monitoring to allow more effective HAB events and other pre-event detection of marine biotoxins. Data-sharing collaboration on sea-surface temperature, salinity, nutrient levels, and other environmental conditions can be utilized in predicting toxin production and spread for the protection of public health.

Chapter 10

Seafood Waste Utilization

Contents

INTRODUCTION

The seafood industry forms a significant part of the world's economy in terms of value to nutrition and the millions of people depending on it for their livelihood. Nevertheless, this industry is responsible for large amounts of waste which mean huge environmental and economic problems. The broad spectrum of seafood waste ranges from fish heads, bones, and skins to shells and internal organs, most of which find their way into landfills. Improperly disposed of wastes can result in water pollution and the production of greenhouse gases, thus having a negative effect on ecosystems and climate change. Moreover, the economic value of these by-products is usually not realized, thus causing the loss of useful resources (Figure 10.1).

Figure 10.1. *Seafood waste.*

Source: Castillo-Téllez B, Castillo Téllez M, Campo MFMd, González EOZ, Domínguez Niño A, Mejía-Pérez GA. Solar-Dried Biofertilizers from Marine Waste: Enhancing the Circular Economy. Sustainability. 2024; 16(15):6593. https://doi.org/10.3390/su16156593.

Recently, there has been increasing interest in the use of seafood waste as a way to address the outlined challenges. Value-added products obtained from waste offer a reduced environmental footprint and open up new economic opportunities for the industry. The emerging biotechniques and innovative processing methods for seafood by-products play a vital role in changing the trend of their use towards the extraction of bioactive compounds, generation of nutraceuticals, and production of food and feed additives. Such efforts not only help reduce waste but also foster sustainability through full resource use and low environmental impact.

Significantly, the use of seafood waste will also be affected by the economics of processing the wastes into saleable products in the market, defined by costs and benefits incurred during the process. Legislation and policy will, in general, be the driving factor in orienting waste management toward sustainability. This will dictate what types of practices are adopted, encouraged, and supported, and the nature of incentives given to companies innovating in this area. The future for seafood waste utilization will thus grow increasingly bound with emerging technologies, interdisciplinary collaboration, and continued research as the industry trends toward more sustainable practices. The current state of the art on seafood waste valorization ranges from the current challenges and practice to its potential for future innovations, highlighting the urgency to introduce a sustainable management dimension into the seafood industry.

10.1. CHALLENGES OF SEAFOOD WASTE MANAGEMENT

Seafood waste management presents some of the most serious environmental and economic concerns. Global demand for seafood continues to rise; its production, consequently, has increased and, directly linked to this, more wastes. This increasing volume of wastes must be properly managed to minimize its impact on the environment and to maximize resource usage and provide a long-lasting livelihood for people working in the industry. There are, however, several challenges to managing seafood waste that are yet to be surmounted.

10.1.1. Environmental Concerns

One of the major environmental problems related to seafood waste is that it easily causes environmental pollution. Seafood waste, if not handled properly, may potentially cause water, air, and natural ecosystem pollution (Arvanitoyannis and Kassaveti, 2008). For instance, if the organic seafood processing wastes are directly discharged into rivers, lakes, or oceans without treating them well enough, they can potentially cause water pollution. In the process of decomposition, organic matter can consume all the available oxygen in water bodies and turn them into hypoxia or dead zones, making it impossible to support aquatic life. This does not affect biodiversity alone but the balance of whole ecosystems.

In addition to water pollution, greenhouse gases, mainly methane and carbon dioxide, are released into the atmosphere through the decomposition process of seafood wastes. These gases are major contributors not only to global warming and climate change but also enhance the negative effects on the planet. The proportion of methane gas emanating from seafood waste going to landfills is huge due to anaerobic breakdowns of organic matter. The carbon footprint of the seafood industry can be reduced by reducing the proportion of seafood waste going into landfills.

Moreover, fish waste can attract pests and scavengers, providing the origin of other environmental problems. Wastes dumped in the marine coastal area or at the back of processing plants can be attractive to rats, birds, and other insects, some of which are known as disease carriers or cause imbalances in the local ecosystem. This is more serious in areas with inadequate infrastructure for managing waste or where poor regulation is implemented.

10.1.2. Economic Concerns

There are also huge economic challenges associated with the handling of seafood waste. Handling seafood waste can be expensive, especially for producers in countries and areas where tight environmental regulations have been put in place. The collection and processing of waste materials for composting or rendering or other processes involves additional operational costs. Sometimes, these costs weigh heavily on SMEs, which may not have the financial muscle to invest in proper measures of solid waste management.

The underutilization of seafood by-products also represents an economic opportunity lost, beyond the ecological impacts. Most of these by-products individually hold many valuable compounds that could be valorized for several industries such as pharmaceuticals, cosmetics, and agriculture. More often than not, they are wasted instead of being valorized due to a lack of infrastructure and technology in processing most of these by-products. This can not only be an economic loss but also contributes to the inefficiency of the seafood supply chain altogether.

The market for seafood by-products is still relatively nascent, and there are several supply chain logistics, product standardization, and market demand challenges related to it. For example, seafood waste can logically be tough to collect and treat into value-added products, mostly in remote or less-developed areas. The lack of standardization of processes and quality controls for by-products makes the establishment of dependable markets a challenge. From an economic perspective, how these challenges are overcome will determine the realization of the full potential of seafood by-products and the creation of waste management systems that are more sustainable and profitable.

10.1.3. Current Waste Management Practices

The existing seafood waste management practices are very varied and region-specific, processed seafood types, and the regulatory environment. In many instances, seafood wastes are landfilled, incinerated, or marine-disposed—each with serious environmental implications. Landfilling, in addition to contributing to methane emissions as mentioned earlier, occupies a large area of land. Incinerating it reduces the volume of the waste, and if not properly controlled, can result in harmful pollutants being released into the air. Ocean dumping, although increasingly uncommon due to international regulations in this area, can result in severe marine pollution that disrupts local fisheries.

Some alternative ways that have been trialed in some areas to provide a more sustainable management of the wastes are composting, rendering, and anaerobic digestion. Composting entails breaking down the organic wastes to produce soil amendments rich in nutrients for use in agriculture. Animal by-products are rendered, some of which are ground up and given back to the animals for feeding. Organic waste transformation is followed by anaerobic digestion, which produces biogas, acting as a renewable energy source. These practices decrease the environmental impact of seafood by-products and form useful by-products (Figure 10.2).

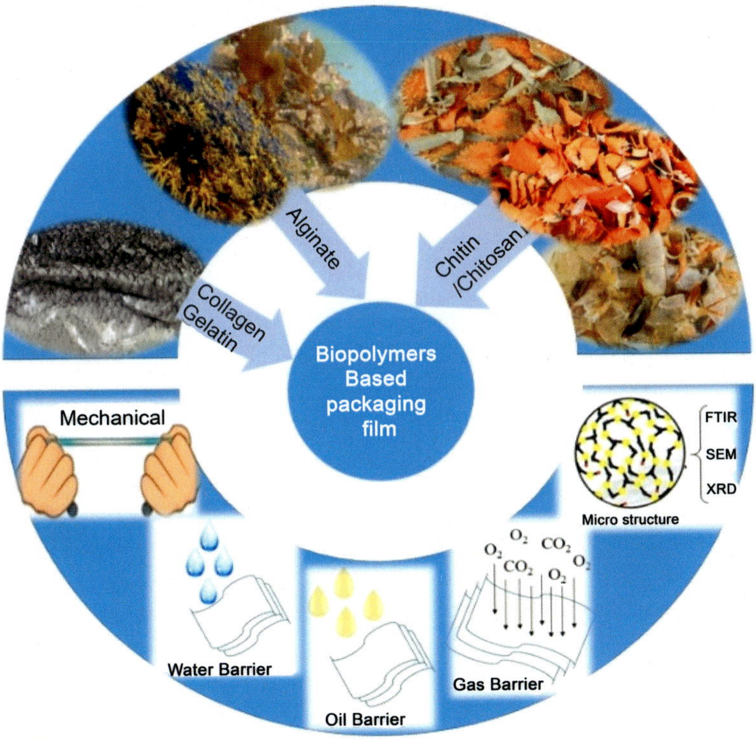

Figure 10.2. *Upgrading of seafood waste as a carbon source.*

Source: Zhan Z, Feng Y, Zhao J, Qiao M, Jin Q. Valorization of Seafood Waste for Food Packaging Development. Foods. 2024; 13(13):2122. https://doi.org/10.3390/foods13132122.

Yet, there are many major barriers to this diffused application: cost, lack of infrastructure, and regulatory barriers. Indeed, in most developing nations, infrastructure for processing wastes is very basic, and less-sustainable techniques of disposal are still prevalent. Besides, expenses related to the operation of developed systems of waste management can be really high, especially in the case of small-scale producers. Moreover, the regulatory frameworks, in some places, may not be effective enough to promote sustainable practices; hence, their implementation is not consistent, and the damage caused to the environment is not curbed.

10.2. VALORIZATION OF SEAFOOD BY-PRODUCTS

Valorization of seafood by-products is becoming increasingly important in the seafood industry, providing value to what were formerly wastes. In this line, it addresses environmental concerns related to waste disposal and opens huge economic potential, therefore contributing to seafood sector sustainability. Fully utilizing by-products to minimize waste while generating added value is increasingly necessary in the face of rising demand for sustainable practices in the seafood sector.

10.2.1. Value-Added Products from Waste

Traditionally, discards or low-value applications such as animal feed or fertilizer were made out of seafood by-products like fish heads, skins, bones, shells, and internal organs. Improved technology and an understanding of their nutritional and functional properties have currently changed this.

Extraction of bioactive compounds from seafood by-products is one of the main recent developments in by-product valorization (Shahidi, Varatharajan, Peng, and Senadheera, 2019). Collagen, gelatin, omega-3 fatty acids, chitin, and chitosan are examples of these bioactive compounds with value-added applications in the food, pharmaceutical, and cosmetics sectors. For instance, collagen and gelatin, originating from fish skins and bones, find wide applications in food supplements, functional food, and beauty products industries due to their beneficial effects on skin, joints, and general health. Another example is omega-3 fatty acids, which are extracted from fish oils and have a very strong reputation regarding their cardiovascular and anti-inflammatory benefits, boosting their demand in the nutraceutical market.

Chitin and chitosan are two such organisms derived from the exoskeletons of crustaceans like shrimp and crabs, which have applications running from food additives and packaging material to biomedical devices and even water treatment solutions. These biopolymers are characterized by their biocompatibility and biodegradability, coupled with antimicrobial properties that make them very useful in different industries.

Other than bioactive compounds, seafood by-products are used in the generation of protein hydrolysates, representing protein-rich powders or liquids used in food and feed formulations. As these hydrolysates represent the enzymatic breakdown of proteins in fish wastes, they become easily digestible products with high nutritional value. Protein hydrolysates have very special functionality in aquafeeds, as they stimulate growth and promote health in farmed fish. They are employed in the food industry to improve the protein content of such end products as soups, sauces, and sports nutrition supplements.

Another very promising area of valorization is the production of fish meal and fish oil, which are both traditional by-products from the seafood industry but have seen increased demand over the last few years due to usage in aquaculture and livestock feeds. Fish meal is a ground, dried remainder of fish, highly proteinaceous, and used as a feed ingredient supporting the growth of farmed fish and animals. Adding fish oil to both animal feeds and human dietary supplements capitalizes on its value as a great source of omega-3 fatty acids. Production of fishmeal and fish oil from by-products rather than from whole fish reduces pressure on wild fish stocks and supports more sustainable fisheries.

10.2.2. Economic Potential and Sustainability

There is huge economic potential in valorizing seafood by-products. The seafood industry can develop new revenue streams and increase its profitability by turning waste into valuable products. This becomes very useful in a highly competitive global context where the margins may be tight and demand for sustainability is growing.

Valorization, in this sense, offers entry into new markets and opens up new business opportunities to a lot of players. As consumers are increasingly being made health- and environmentally conscious, there is an augmented demand for products formulated with natural and sustainable ingredients. In this context, seafood co-products will strongly emerge due to their high nutritional value and functionality. Companies willing to invest in development and commercialization can capture these emerging markets and deliver sustainable competitive advantages.

Valorization of seafood by-products can contribute to the overall sustainability of the seafood sector. If waste is reduced and resources are used to their fullest potential, environmental impact will be minimized and business will shift towards a circular economy. This would not only help conserve marine ecosystems but also support the achievement of global sustainability goals as outlined in the United Nations SDGs, which call for responsible consumption and production.

Valorization of by-products will bring added economic value to companies and the economy as a whole. New products and markets create jobs and stimulate economic activity, in particular within coastal communities, where the seafood industry normally is an important local economic driver. Further, innovation in new processing technologies and new uses for by-products may result in further investment in research and development, thus driving further growth and sustainability.

Despite all these advantages, a variety of challenges has to be resolved before the full economic potential of seafood by-product valorization is realized. They involve infrastructural and technological investment, the development of standardized processes for processing by-products, and the establishment of quality control measures that would ascertain the safety and efficacy of value-added products. Closer collaboration on these lines among the seafood industry, research institutions, and government agencies is therefore important for addressing these challenges and putting in place an enabling environment for by-product valorization.

Regulatory support is also key to the long-term use of seafood by-products. Governments can drive the use of by-products through policy support, funding of research and development, and creating market conditions that favor value-added products. It is also important to establish a clear processing and labeling framework for by-products, which will help create trust among consumers in the market success of these products.

10.3. EXTRACTION OF BIOACTIVE COMPOUNDS FROM SEAFOOD WASTE

Seafood wastes can be used as a source of extraction of bioactive compounds. In this case, they can effectively be considered as a by-product resource for both nutritive and functional purposes. The huge potential of these bioactive compounds i.e., proteins, peptides, lipids, and polysaccharides, remains underexploited in terms of pharmaceutical, nutraceutical, cosmetic, and food applications. The challenge lies in efficiently extracting these compounds while maintaining their bioactivity, purity, and functionality.

10.3.1. Types of Bioactive Compounds

The waste from seafood, including heads, skins, bones, scales, and shells of fish and crustaceans, is a treasure trove of several bioactive compounds that have various health benefits and industrial applications. Among these, proteins and peptides are considered quite valuable. For example, proteins obtained from fish skins and scales have the potential to release bioactive peptides after hydrolysis, exerting antioxidant, anti-hypertensive, and anti-microbial activities, among other biological functions. These peptides are gaining importance in nutraceuticals and functional foods for the cardiovascular system, for improving immune function, and also as antioxidants.

Lipids are another important group of bioactive compounds extracted from wasted seafood. In particular, omega-3 fatty acids, extracted from seafood, offer many health benefits. Fish-derived oil from the fatty fish tissues and co-products has already been well recognized as a source of several important bioactive compounds, including eicosapentaenoic acid and docosahexaenoic acid. These fatty acids maintain cardiovascular health; support brain functioning; and have anti-inflammatory functions, in addition to possessing these potential health benefits. Hence, the process of extracting high-purity omega-3s from seafood wastes contributes not only to the value addition of these by-products but also to the sustainability of the production of omega-3 while it reduces the need to harvest extra fish.

Another class of bioactive compounds commonly extracted from seafood waste, especially from the exoskeletons of crustaceans like shrimp and crabs, includes polysaccharides such as chitin and chitosan. Chitin is a naturally occurring polymer that can be converted into chitosan through chemical or enzymatic means, after which it finds very useful applications on account of its biocompatibility, biodegradability, and antimicrobial properties. Chitosan finds wide use in different fields, such as for wound dressings, water treatment, food preservation, and as a carrier for drug delivery systems.

Besides these compounds, seafood wastes also become a source of important minerals like calcium and phosphorous, which are extracted from fish bones and scales. These are truly essential minerals for bone health and utilized in dietary supplements or fortified foods. Other bioactive compounds from seafood waste include carotenoids, collagen, and glycosaminoglycans connected with several industries that range from cosmetics to pharmaceutical uses.

10.3.2. Techniques for Extraction

Bioactive compounds extracted from seafood wastes, call for special techniques to achieve efficient, pure, and bioactive products. A variety of techniques, starting with traditional methods such as solvent extraction and enzymatic hydrolysis to advanced technologies like supercritical fluid extraction, membrane filtration, and ultrasound-assisted extraction, are employed.

- **Solvent Extraction:** This is the method most applied in the extraction of lipid components and other hydrophobic compounds from seafood waste. This is a process that utilizes organic solvents acting in breaking, linking, or dissolving the waste matrix from the compound of interest. This technique has, therefore, worked well to get fish oils and omega-3 fatty acids, but sometimes it demands further steps of solvent removal and purification of the end product. The conditions of extraction and solvents should be carefully optimized for maximal yield, preserving the bioactivity of the extracted compounds.
- **Enzymatic Hydrolysis:** It is one of the more popular techniques for extracting proteins and peptides from seafood waste. Specific enzymes, such as proteases, are used to cleave proteins into smaller peptides and amino acids. Enzymatic hydrolysis is preferred because of its mild conditions, which assist in the preservation of the bioactivity of the peptides. The selection of the enzyme and control of hydrolysis parameters, such as pH, temperature, and time, becomes very critical to modulate the extent of bioactive hydrolysis and the functionality of the resulting peptides. By this technique, bioactive peptides can often be generated, having specific health benefits including antioxidant or antihypertensive activities.
- **Supercritical Fluid Extraction (SFE):** This is an advanced extraction technique in which supercritical fluids, especially carbon dioxide, can be used in the extraction process of the bioactive compounds from seafood waste. Such an intermediate of carbon dioxide, possessing properties of both a liquid—having a solvating power—and a gas—having diffusivity—in the supercritical state, makes it possible to extract such compounds as lipids, carotenoids, and other non-polar. SFE is sometimes considered the green alternative to solvent extraction because it eliminates the use of organic solvents and ensures proper control over extraction conditions. The high efficiency and selectivity of SFE position ideally this process for the production of high-purity omega-3 fatty acids and other valuable lipids from fish waste.
- **Membrane Filtration:** Ultrafiltration and nanofiltration techniques are used to separate bioactive peptides and bioactive proteins in seafood waste. These are processes where a semi-permeable membrane is used through which there is only a certain degree of passage of molecules based on their size or molecular weight. Simple and fractional membrane filtrations are commonly combined

with enzymatic hydrolysis for the purification and concentration of bioactive peptides. The main advantages caused by using membrane filtration are the high selectivity, correct scalability, and operation under mild conditions, which enable the bioactivity of compounds to be retained.

- **Ultrasound-Assisted Extraction (UAE):** This is a new technique that uses ultrasonic waves to enhance the extraction of bioactive compounds from seafood wastes. The mechanical effects of ultrasound, such as cavitation and agitation, enhance the penetration of solvent within waste material and the mass transfer of target compounds (Naghdi et al., 2023). Particularly, UAE has shown excellent ability in the extraction of polysaccharides, including chitin and chitosan, from the shells of crustaceans. This is an appreciated technique that allows reduction of extraction time, lowering of solvent consumption, and raising the yield in comparison to conventional techniques.
- **Fermentation:** It is yet another way of deriving bioactive compounds from seafood waste with increased functional properties. This is the use of mainly bacteria and fungi, which are responsible for the cleavage of the very complex molecules that are needed to convert them into simpler bioactive compounds. Fermentation can be aimed at the production of bioactive peptides, omega-3 fatty acids, and other compounds with considerably improved functional properties. The importance is also noted in the selection of microbial strains and optimization of fermentation conditions for high yields and desired bioactivities.
- **Microwave-Assisted Extraction (MAE):** The extraction solvent and the material to be extracted are heated by microwave energy, thereby enhancing solubility and the kinetics of extraction of the solvent into the material. In the literature, the use of this technique in the extraction of lipids and other bioactive, thermostable compounds from seafood waste has been well demonstrated. Automation in MAE reduces the time of extraction, reduces the volume of solvent, and provides high-purity extractables.

10.4. APPLICATIONS OF SEAFOOD WASTE-DERIVED PRODUCTS

Interest in seafood waste utilization has grown in recent years because seafood waste disposal has become an environmental issue and that such by-products may be economically transformed into useful products. Valuable seafood wastes, such as fish heads, skins, bones, and shells, are important sources of bioactive compounds applied in industries related to the production of nutraceuticals, pharmaceuticals, food additives, and animal feed.

One of the most prominent applications of seafood waste-derived products is in the nutraceutical and pharmaceutical industries. Nutraceuticals are derived from food sources that can offer health benefits beyond basic nutrition, normally used as dietary supplements. Seafood waste is a rich source of bioactive peptides, omega-3 fatty acids, and other important nutrients that can be extracted and formulated into nutraceuticals. For instance, fish oil, which is a rich source of omega-3 fatty acids, such as EPA and DHA, is commonly used in supplements for maintaining heart health, and brain function, and lowering inflammation. Other than this, collagen and gelatin, which are extracted from the skins and bones of fish, are normally used in dietary supplements for joint health, skin elasticity, and wound healing. These compounds not only add value to health benefits but also provide value addition to seafood by-products that otherwise become waste.

Seafood waste bioactive compounds are currently under investigation in the pharmaceutical industry for their prospective therapeutic applications. Chitosan, a derivative of chitin, found in the shells of crustaceans, has been getting attention for its biocompatibility and antimicrobial properties, proving useful in wound dressings and drug delivery systems or tissue engineering. More than that, bioactive peptides from fish proteins proved to have huge potential for treating hypertension, diabetes, and obesity due to the capability to modulate biological phenomena in the organism. Their interest in these compounds arises out of the need for natural and renewable sources of therapeutic agents; seafood waste offers an abundant and cost-effective supply.

Products resulting from seafood waste also serve important roles in both the food and feed industries. In the food industry, these by-products are used for developing additives that enhance the nutritional and shelf-life properties of various foodstuffs. For instance, fish protein hydrolysates—a resultant product of enzymatic hydrolysis of fish proteins—are used as flavor and flavor enhancers, besides being functional ingredients in soups, sauces, and processed foods. Hydrolysates are rich in amino acids and peptides, hence acting as food property enhancers in terms of taste, texture, and nutritional value. Added to this, omega-3 fatty acids extracted from fish oil are added to a wide range of foodstuffs, such as margarine, yogurt, and even baby formula, to 'medicalize' their health values.

In the animal feed industries, seafood waste-derived products play an important role as quality feed additives for livestock, poultry, and aquaculture.

Fishmeal is important in feed for its high protein content and essential amino acids, made from the dried and finely ground remains of fish (Wadhwa and Bakshi, 2016). In particular, it forms an important protein source for farmed fish and shrimp in aquaculture and will continue to be so in promoting growth and feed efficiency. Chitin and chitosan, derived from crustacean shells, have been used not only as feed additives in the overall growth performance enhancement of animals but also in animal immunity enhancement and gut health improvement. Such applications give seafood waste meaningful and sustainable uses while reducing the environmental impact related to the production of traditional feed.

10.5. ENVIRONMENTAL IMPACTS OF SEAFOOD WASTE

The improper handling of seafood waste can result in pollution, alter ecosystems, and further exacerbate an already overstressed situation of marine resource depletion. Understanding the impact and devising proper ways to mitigate it would be important for the induction of real sustainability within the seafood industry. When the waste settles on the seafloor or begins decomposing on land close to coastal areas, it may release deadly ammonia, methane, and hydrogen sulfide gases. These substances dissolve in water, the water being depleted in oxygen, forming so-called "dead zones" in which marine life cannot exist due to the low supply of oxygen in the water. This odor created by the decaying material further reduces the quality of life for local residents and tarnishes the tourism potential.

The accumulation of seafood waste will also disturb the community's local ecology. More specifically, the input of greatly enhanced nutrient levels into a marine system can place the natural system out of balance with nutrients, promoting the overgrowth of some forms of life, such as algae, to the detriment of other creatures. This eutrophication process may lead to algal blooms of toxic algal species that shade sunlight and use up oxygen in their respiration, and in the final end deprives the water of oxygen, leading to fish kills and kills of other aquatic animals. This waste further affects the feeding behavior of scavenging species through the alteration of predator-prey relationships, thereby disturbing balance among the species.

It is, therefore, crucial to develop ways to minimize the generation of waste and its disposal such that these environmental impacts can be put to rest. One of the most effective approaches is actually reducing the amount of waste produced at the source. This can be achieved through better fishing practices that are friendly to the environment and reduce bycatch together with the innovation of technologies in the processing that leaves no part behind. With huge maximization of all the parts of the fish, less or no waste is generated, and dropout becomes a lesser requirement.

Another important strategy is converting seafood waste into valuable products, therefore categorized under valorization. For instance, the development of new uses for by-products, such as fish bones developed for fertilizers or fish skins for collagen supplements, significantly reduces the portion thrown as waste. This would reduce the environmental impacts and at the same time open up economic opportunities in business for all those operating seafood ventures.

Proper waste management practices are also crucial in the prevention of pollution and safeguarding of the ecosystem. Waste should be collected and processed in facilities meant to handle organic material safely. For example, seafood wastes could be composted and converted into a good, rich soil amender and applied to the farm. In this way, it saves the volume of the waste and recycles valuable nutrients back into the environment.

There are strict regulations and monitoring channels to ensure that garbage disposal practices do not have any detrimental impact on the environment. Governments and industry bodies can put in place set standards on how to dispose of seafood waste. For example, there can be a total ban on disposing of waste in the ocean and insistence on treating the waste before bagging the waste for disposal. There is a need for effective enforcement to ensure that the practice is adhered to in a bid to save the marine ecology.

In addition, education and awareness campaigns can therefore reduce the profile of seafood wastage through more regulatory measures. Stakeholder awareness of the impacts caused by poor waste management, on the one hand, and the advantages of sustainable operations, on the other, will go a great way toward influencing the behavior of stakeholders across the seafood value chain to one of increased responsibility. However, the consumers may be sensitized to the need to support sustainable seafood. This will create a market with a demand for products that in turn cause minimal waste and environmental impact.

10.6. BIOTECHNOLOGICAL APPROACHES FOR SEAFOOD WASTE UTILIZATION

Biotechnology is a hopeful solution for the proper utilization of seafood wastes, which always seem to be a burden and actually can turn out to be valuable resources. Innovative processing methods and sophisticated biotechnological approaches make it possible to get useful products from seafood by-products, contributing to both environmental sustainability and economic growth.

One of the major benefits of biotechnology in regard to seafood waste utilization is its ability to extract several important bioactive compounds. These compounds, among them proteins, peptides, oils, and enzymes, are usually present in parts of seafood normally considered waste—skins, bones, and internal organs. Through biotechnological processes, these by-products can be reprocessed into high-added-value products for different industries. For instance, fish skins can be turned into collagen, a material broadly applied in cosmetics and pharmaceuticals. Similarly, it is possible to extract fish oils rich in omega-3 fatty acids for use in dietary supplements.

Finally, value addition in processing seafood waste involves innovative methods of breaking down the waste into smaller manageable components. One of the important techniques used in breaking down seafood waste is enzymatic hydrolysis. Enzymatic hydrolysis is a process where specific enzymes are used to decompose proteins and other organic materials into simpler molecules like peptides and amino acids. These more easily extracted smaller molecules can contribute to functional foods and animal feed formulations, and even to bioactive peptides with specific human health benefits. The method is highly efficient, and also from an environmental point of view, it is friendly, since it has a very low dependence on aggressive chemicals and very low production of wastes.

Another biotechnological approach that has huge potential for the utilization of seafood wastes is fermentation. Microorganisms such as bacteria and fungi can raise fermentation of seafood wastes into useful goods, including biofertilizers, animal feed, and even biofuels. For example, fish wastes could be fermented to produce fish silage that would be a high-protein feed in both livestock and aquaculture. Thus, the fermentation process not only makes use of waste otherwise disposed of but also provides sustenance to other related industries.

Biotechnology, too, has various solutions to the environmental challenges related to seafood waste. One of the biotechnological approaches is bioremediation using microorganisms that degrade and remove marine seafood wastes from the environment, hence reducing impacts (Venugopal, 2021). This will decompose hazardous material inclusive of organic pollutants and heavy metals in such waste, hence making it safe for disposal or use. It is not only going to help manage the waste better, but it will also prevent pollution and contribute to environmental protection.

Among them, the newer processes for seafood waste utilization include the biotechnological applications for developing new products. Advanced genetic engineering that could genetically manipulate microorganisms can let them produce certain enzymes or compounds useful for various industries. For example, the gene coding for chitinase—an enzyme breaking down chitin, the main constituent of crab shell material, into chitosan—can be engineered into bacteria. Chitosan is an important biopolymer used in medicine, agriculture, and preservation of foodstuffs; therefore, it is among the most demanded products in the market.

Furthermore, biotechnology makes it possible to produce biofuels from seafood wastes in an environmentally sustainable way. Anaerobic digestion converts the by-products of seafood into biogas, which is a renewable energy source that could be employed as a heat source, for the production of electricity, or even as biofuel in vehicles. It will provide not only a substitute for energy sources but also help to reduce the carbon footprint of the seafood industries by being prepared out of wastes that would otherwise turn into a source of GHG.

Contributions of biotechnology to seafood waste utilization go beyond processing and extraction. It also contributes to the attesting of sustainable practices that guarantee efficient utilization of the resources available. Integrated biotechnological systems, for instance, can be designed in such a way that every part of seafood is utilized to make sure minimal wastage occurs while maximizing the value attained from each catch. This approach goes in line with the principles of circular economy, where waste is minimized and resources are always reused.

10.7. ECONOMIC ASPECTS OF SEAFOOD WASTE UTILIZATION

Economical aspects of the utilization of seafood waste have been gaining the spotlight because industries and researchers realize that wastes are wealth yet to be exploited. Conventionally, seafood waste was a by-product that had no value or low value, usually resulting in environmental problems while at the same time costing money for its disposal. Nevertheless, seafood waste, given advances in technology and with growing emphasis on sustainability, is now seen as a resource that has enormous potential for contributing to economic growth while reducing environmental impacts.

Cost-benefit analysis is naturally a very key factor in the economic evaluation of the utilization of seafood wastes. This would involve estimating the costs of collection, processing, and conversion against the potential revenue to be generated from such products. The initial investments in technology, infrastructure, and research can be huge, but often long-term benefits outweigh these costs. Through the valorization of waste into products like fish meal, production of bioactive compounds, or biofuels, a company can realize new income streams that help recover the upfront investments and deliver considerable profits. Moreover, valorization of seafood waste will help save money spent on handling and disposing of the waste. Close to what has been argued, instead of spending money on waste removal and processing for dumping, companies can use their funds for processing and valorization of the waste, thus closing the cycle by constantly reusing and recycling it. Additionally, this will be an approach toward greening the seafood industry while setting an example of handling and minimizing most of the costs involved in waste disposal.

The potential for seafood waste-based products is huge in the market. For instance, fishmeal and fish oil, prepared from processing wastes, have already been under huge demand by the aquaculture and livestock industries as higher value-added, high-protein feed ingredients. Another example is that bioactive compounds produced from seafood waste have huge potential use in the pharmaceutical, cosmetics, and nutraceutical industries. It also makes them more viable in the market as waste-derived products, hence quite attractive as business investments for businesses looking at consumer demand for natural and sustainable products.

Biofuel production from fish wastes is yet another area that holds so much promise. As the world limits its reliance on fossil fuels, shifting its sources of energy to renewable ones, biofuels become highly relevant (Yuvaraj et al., 2016). Companies could transform seafood waste into either biogas or biodiesel, hence helping them penetrate the increasing renewable energy market for an additional source of income while contributing to environmental sustainability.

10.8. REGULATIONS AND POLICIES ON SEAFOOD WASTE MANAGEMENT

Seafood waste regulations and policies must be put into place to enable the industry to handle its activities sustainably and responsibly. Such frameworks prop up the guidelines that involve the development, conveying, and running of seafood ways that are environmentally friendly and maximize available resources. Managing seafood properly is not only an environmental necessity but also an economic opportunity for creating value from what would otherwise have been a waste stream.

International and national regulatory frameworks on seafood waste are established to address the environmental and public health challenges related to seafood waste. The regulations would allow for setting standards of waste disposal, hence protecting water bodies and the environment around the water bodies from hazardous substances. For instance, pollution in coastal areas from fish offal and other by-products has already been controlled by ground guidelines in many countries. "28 such regulations charge that seafood processors treat waste before its disposal, with the effects of minimizing environmental harm." Additionally, traceability and monitoring of seafood waste is a procedure included in the many regulations. This therefore calls for a way of tracing the origin and the destination to which the waste products from seafood products go, in a manner that will guarantee proper treatment. Such measures are often aimed at preventing any form of illegal discharge and seeing to it that the waste is treated at facilities prepared to manage it safely. In most regions, compliance with these regulations is a must, with penalties for violations either being monetary or the suspension of operating licenses.

Besides regulatory frameworks, there are a number of incentives for sustainable seafood waste management. Many governments and international organizations provide financial support in the form of subsidies or tax breaks to companies with an interest in contributing to more sustainable waste management. They may come in the form of monetary encouragement to enable research and development in waste-processing technologies, grants for companies that may be working on the waste-to-resource conversion process, and even reduced taxes for businesses that attain preset environmental standards. Such incentives are hence offered by policymakers to reduce the impact of the seafood industry on the environment, in the view of providing the required impetus for innovation and economic growth.

An example is businesses that invest in technologies to convert byproducts derived from seafood into biofuels, fertilizers, or any other detention value; this is by the government offset of initial project costs. These moves should not just leave a project to stand on its own but should open up avenues for more mainstreaming such practices across the industry. For instance, some of the programs financially reward companies for cutting down on overall waste generated, further incentivizing their practice to avoid generating waste and to be effective in resource use.

Implementation of food waste management policies also forms another critical part of seafood waste management. This is primarily achieved through the establishment of closed-loop systems in which the by-products of waste are always reprocessed and reused rather than disposed of indiscriminately. Circular economy policies are now gaining a lot of support from governments and regulatory authorities mainly because the models do not only reduce the environmental impacts emanating from seafood waste but also introduce new economic potentials. For example, policies have the potential to include the added value of using byproducts from seafood in animal feeds, fertilizers, and new material development, thus pointing out a potential market for products derived from its wastes.

10.9. FUTURE PROSPECTS IN SEAFOOD WASTE UTILIZATION

Prospects for the future use of seafood wastes are promising with the development of some emerging technologies, innovations in this area, and interdisciplinary collaboration toward the zero waste goal. As the seafood industry grows, the attention it gives to waste reduction and turning by-products into valuable resources remains significant.

Modern extraction technologies, such as advanced biotechnological methods, hold the potential to extract bioactive molecules with novel functionalities from seafood waste. Recently developed technologies such as enzyme-assisted extraction and fermentation, and green chemistry approaches, are more efficient and environmentally friendly. Product innovations in the area could help develop new products in nutraceuticals, pharmaceuticals, and also in sustainable packaging, further enhancing the economic potential of seafood waste utilization.

Collaboration among industry, academia, and government can be used to drive research and development in the field. Such collaborations can create new markets for waste-derived products and lead the way for introducing more sustainable practices into the industry. Important issues will be solved about regulation so that new products can be developed with the proper safety standards and accepted by the consumer.

That amount of latitude in future possibilities opens up improvement, sustainability, and economic soundness for the utilization of seafood waste. Finally, with time, research progress and the development of new technologies might enlarge the potential transformation from waste to valuable resources in setting the ground for more sustainable and efficient seafood preparation and processing industries.

Chapter 11

Advances in Seafood Processing Technologies

Contents

INTRODUCTION

Over the recent years, the seafood industry has advanced amazingly fast. This is accredited to progress in technologies associated with the processing sector regarding quality, safety, and sustainable production of seafood products. Innovation is sought after as the challenges, on one hand, come from the high perishability of seafood and, on the other hand, the fast-growing world demand for fresh and high-quality seafood.

Some of the more exciting developments involve new technologies such as high-pressure processing (HPP), pulsed electric field (PEF) technology, and superchilling. Each of these techniques offers unique advantages and has the potential to transform the seafood industry in meaningful ways.

HPP is a breakthrough technique that uses extremely high pressure and kills harmful bacteria and spoilage organisms in seafood without the reliance on heat (Campus, 2010). This is an important finding because many heat-based methods can distort seafood's taste, texture, and appearance. With HPP, seafood products retain the same fresh and natural properties of their source while becoming safe and shelf-stable. This can be a benefit for certain very perishable seafood items, such as raw oysters, sushi-grade finfish, and other ready-to-eat meals where freshness is of critical importance. In addition to preservation, HPP has been advanced to recover rich nutrients from seafood by-products, bringing new opportunities to create high-value products from what would otherwise be considered waste.

PEF is another promising technology to emerge, with the application of short bursts of high voltage introduced for applying to seafood. This way, the cell membranes of microorganisms are effectively killed by forming small pores in them (Gómez et al., 2019). This not only stops the activities of the microorganisms but also enhances the tenderness and juiciness of the seafood. PEF is particularly well suited for seafood that is sensitive to heat, like shrimp, squid, and scallops, where typical thermal damage from standard methodologies would occur. Moreover, PEF can render seafood more susceptible to marinating and flavoring agents, and value-added products respond to the need to deliver products perceived as healthier and more appealing to the consumer.

Superchilling, also called partial freezing, is another novel approach developed in the seafood industry. It consists of the elevation of seafood temperature and slight lowering below freezing, in the range of -1°C to -2°C, in the absence of ice crystal formation (Magnussen, Haugland, Hemmingsen, Johansen, and Nordtvedt, 2008). This makes superchilling a very potent technique to dramatically retard the natural processes that lead to spoilages to extend the shelf life of seafood, yet maintain the fresh-like quality. Unlike in normal freezing, the texture of the seafood does not worsen after it is defrosted; therefore, it maintains the structure of the seafood. This makes it an ideal choice for high-end fish products, such as sashimi grade, and other luxury products. Superchilling, when applied with modified atmosphere packaging (MAP), can enable an enhanced preservation environment and prolong the freshness of seafood.

11.1. INNOVATIONS IN SEAFOOD QUALITY ASSESSMENT

While traditional techniques of quality assessment by sensory evaluation and simple chemical analysis have served the industry well in the past, they are often labor-intensive, subjective, and have limited scope. New technologies and tools move the industry toward more accurate, efficient, and inclusive methods for seafood quality evaluation. These innovations set the bar in raising quality control precision but also lay a foundation for the advancement of the field in the future.

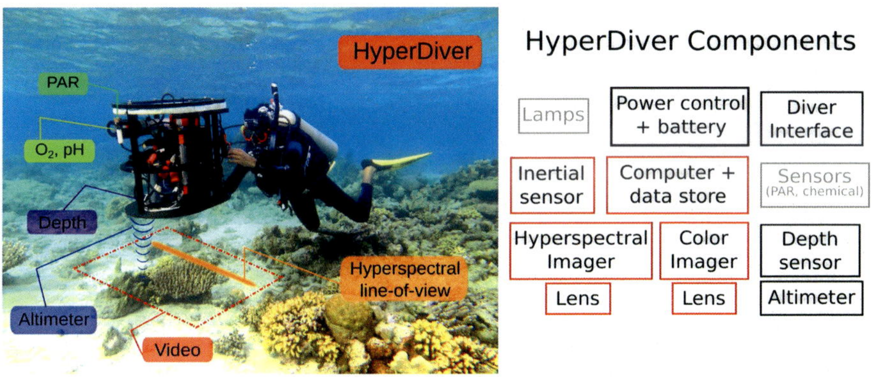

Figure 11.1. *A diver-operated hyperspectral imaging and topographic surveying system for automated mapping of benthic habitats.*

Source: Chennu, A., Färber, P., De'ath, G. et al. A diver-operated hyperspectral imaging and topographic surveying system for automated mapping of benthic habitats. Sci Rep 7, 7122 (2017). https://doi.org/10.1038/s41598-017-07337-y.

One of the most significant developments in seafood quality assessment is now that hyperspectral imaging has become an extensively applied tool. This technology captures and processes information throughout the electromagnetic spectrum, thus allowing non-invasive analysis of seafood products. Hyperspectral imaging provides detection of very small changes in seafood structure say for instance, changes in moisture content, fat levels, and protein distribution that are very critical pointers of quality (Cheng and Sun, 2014). From there, spectral data analysis enables HSI to identify defects, contamination, and freshness with a very high accuracy level. The technique is quite useful in evaluating large batches of seafood products in a real-time setup. Hence, it is an excellent tool for processors who expect uniformity in their seafood products (Figure 11.1).

Another promising innovation is nuclear magnetic resonance spectroscopy, which performs an exquisitely fine-grained analysis of the molecular constitution of seafood. NMR spectroscopy is highly sensitive to changes in the biochemical composition of seafood, in particular, the degradation of proteins and lipids, which, in turn, may indicate spoilage. Using NMR, it is possible to derive a more precise explanation of the quality and freshness status of seafood than is possible by conventional chemical tests. Moreover, NMR spectroscopy can be applied for seafood species identification, fraud testing, and label control, something increasingly important on the global market, where misdescription and fraud are already quite common.

The other new promising tool in seafood quality control is the application of biosensors. A biosensor is a device using biological molecules, usually enzymes or antibodies, to identify chemical compounds indicating spoilage or contamination. For example, the presence of histamine can be detected through a biosensor; histamine is a compound formed in fish due to poor storage and might make one become food poisoned. Some advantages of biosensors are very fast response time, high sensitivity, and field or process line applicable. With continued refinement and drops in their prices, they are bound to have an important role in ensuring food safety and quality control of catch and aquaculture products.

Besides these technological developments in their quality assessment and traceability chain, there is a thought for the industry to turn to blockchain technology. Blockchain presents a decentralized, immutable ledger, which tracks every step of seafood supply—from catch to consumer. It makes it easier to ensure that seafood products are handled under the best possible conditions in terms of storage and transport by offering a clear and tamper-proof record of each transaction. It is not only enhancing traceability, but this technology also builds up consumer trust through verification of the authenticity and quality of the seafood being purchased. As blockchain becomes more integrated into the industry, this could revolutionize how quality concerning seafood is monitored and maintained across the supply chain.

11.2. NUTRACEUTICALS AND FUNCTIONAL FOODS FROM SEAFOOD

Nutraceuticals and functional foods have been gaining in popularity; the former refers to bioactive compounds of nutritional origin that have medical or health benefits in addition to basic nutrition, while the latter stands for foods fortified with these bioactive ingredients. Seafoods, being rich in several bioactive compounds like omega-3 fatty acids, peptides, and antioxidants, find increasing attention as an ideal raw material for those products. The value addition of seafood towards the development of nutraceutical and functional foods will not only harness the manifold health benefits accrued from these bioactive compounds but also resonate well with market trends favoring natural and healthy product options.

One of the more prominent types of seafood nutraceuticals is omega-3 fatty acids, more precisely, eicosapentaenoic acid, and docosahexaenoic acid. Very well known for offering anti-inflammatory actions and supporting brain function (Jing-Jing Li and Huang, 2008), these are some of the long-chain polyunsaturated fatty acids that have positive benefits on cardiovascular health. These benefits, now increasingly popular with research and an ever-growing body of it in number, have gradually increased consumer interest in, and use of, omega-3 supplements and driven, in turn, other developments, such as fish oil capsulation and fortification. Advances in extraction and purification technologies further resulted in the development of high-purity omega-3 concentrates, which are now being used in different functional food applications, including fortified dairy products, spreads, and beverages. This will ease the incorporation of these beneficial compounds into the consumer's daily diet.

Other bioactive peptides from seafood are receiving increasing attention in this regard, going beyond the health benefits of the omega-3s. For instance, fish skin and scale-derived marine collagen peptides are being used more and more in functional foods and nutraceutical applications for promoting skin, joints, and wound health. The global collagen supplement market is growing at a rapid pace, and the demand for anti-aging and beauty products is a good opportunity where marine collagen presents itself as a value-added ingredient, owing to more bioavailability than bovine or porcine sources of collagen. In the same fashion, fish muscle peptides and their by-products have demonstrated antihypertensive activity, hypocholesterolemic activity, and even antioxidant activity, which open new opportunities for product development in the nutraceutical industry.

Seafood is also a primary source of strong antioxidants, such as astaxanthin, originating from microalgae and marine organisms, for the most part, shrimp and salmon. Astaxanthin is noted to have powerful, potential antioxidant activities, which are likely to benefit health from healthy skin and eyes to decrease oxidative stress related to aging and chronic diseases. This realization has been attributed to the rise of an umbrella of astaxanthin-based products that include dietary supplements, skin care products, and fortified foods. The rising scope of research on health benefits through astaxanthin is likely to increase the speed of innovation in product development, particularly concerning functional foods, where the demand for natural and efficacious ingredients seems to be reasonably high.

11.3. SUSTAINABLE PRACTICES IN SEAFOOD PRODUCTION

With the mounting pressure from environmental challenges, consumer demand, and the quest for the long-term viability of the seafood industry, sustainable practices are emerging. To enhance productivity and reduce their ecological footprint on the environment, sustainable practices in seafood production involve integrating green technologies and eco-friendly methods. These efforts are relatively important, especially in dealing with issues such as overfishing, habitat destruction, and high carbon emissions linked with traditional methods of production and processing of seafood. In as much as sustainability increasingly takes to the center, the role of these practices is accelerating growth within the industry arena, loud not just in environmental stewardship but also as a strategic imperative to competitiveness in markets.

Some key innovation areas in sustainable seafood production include the uptake of green technologies to reduce the aquaculture and wild-catch fisheries impact on the environment. Recycling aquaculture processes is one of the most remarkable developments in sustainable aquaculture. Recirculating aquaculture systems (RAS) technology enables farming fish in a controlled environment where water is filtered and reused continuously, drastically reducing the requirement for freshwater resources and minimizing the discharge of wastes into natural ecosystems (Ahmed and Turchini, 2021). This system also reduces the risk of disease transmission to wild populations and eliminates the need for antibiotics, so commonly used in conventional aquaculture against infections. Aquaculture, by virtue of RAS, is also more productive with significantly reduced environmental risks due to conventional fish farming.

Selectivity in wild-capture fisheries development and use has played an important role in offering sustainability. In traditional methods of fishing, a large amount of bycatch is generated due to fishing operations that is, by nature, the catching of unintended non-target species, such as juveniles, protected species, and marine mammals. To address this issue, modifications have been introduced, among others, using escape-panel-modified nets, circle hooks, and acoustic deterrent devices, all of which have been used to enhance the escape opportunities for non-targeted animals, supporting selectivity. It is such equipment that is thus advanced and has become promoted by regulation and certification programs that encourage the industry to practice responsible fishing, where the sector should be rewarded for reducing its ecologically adverse influence yet getting the highest yields.

Responsible sourcing and processing techniques at the source of seafood also make up sustainable seafood production technologies. The increasing popularization of integrated multi-trophic aquaculture (IMTA) basically shows that the industry is going towards more holistic and sustainable methods. The general concept of IMTA involves the co-cultivation of several species from various trophic levels within the same system, for example, combining fish farming with the simultaneous cultivation of shellfish and seaweed. This approach emulates ecosystem performance in which through natural selection, various species evolved to thrive in the same niche, with all of them contributing to the general well-being of the environment. In an IMTA system, the wastes from the fish can become nutrients for the seaweed and shellfish in a symbiotic relationship, tightening the environmental impact of the given aquaculture operation. The two major advantages are the increment of resource efficiency through added values and production diversification, reducing the reliance on external inputs, like feed and fertilizers.

Sustainability in seafood production is also closely linked with the use of alternative and feed sources in aquaculture. Conventional fish feeds are mainly dependent on products derived from the wild stocks of fish, with species like anchovies and sardines overfished to fill that demand. The industry has answered such a call of responsibility by increasingly turning to alternative protein sources, among which are insect meal, algae, and single-cell protein derived from either bacteria or yeast. They can be produced with lower environmental footprints and without relying on wild fish, making them more sustainable. Besides increasing the inclusion of other alternative feed ingredients, the development of feeds is likely to progress together with the processing technologies to enhance the nutritional quality of the feeds to be sufficient to meet the dietary needs of farmed species without any compromise in growth and product quality.

The importance of sustainability not only resonates in the seafood industry but is driving its growth. Consumers are becoming more realistic with the environment, and hence the demand for seafood that is sourced and produced sustainably is on the rise. The pull being created by the changing consumer preference would lead the industry toward adopting sustainable practices since companies increasingly recognize sustainability not as just a regulatory issue but as a market differentiator. The other significant role that MSC, together with the Aquaculture Stewardship Council, plays for a firm is communicating its commitment to the environment. Being a certification program enables the respective consumers to be assured that the products purchased have, through the process, met the set environmental and social standards, thus increasing loyalty towards the particular brands and consequently increasing market opportunities.

11.4. INTEGRATION OF OMICS TECHNOLOGIES IN SEAFOOD RESEARCH

Integration of omics technologies: genomics, proteomics, and metabolomics in the frame of seafood science, finally realizing the current molecular understanding of marine organisms. Technologies of this kind allow for overall studies of genetic, protein, and metabolic profiles to unravel priceless insights into biology, health, and quality. Further progress in seafood research presumes that these omics approaches open up brand new prospects for seafood production, quality control, and sustainability (Power et al., 2023). But they also raise a new set of challenges in their own right, challenges that have to be addressed if their full potential is to be realized.

Genomics is literally the study of all the DNA in an organism, including its genes. Genomics, therefore, in seafood research is applied in attempts to understand the genetic make-up of the marine species, which has diverse applications. For example, the use of genomics can contribute to the development of genetically improved strains of farmed fish that are disease-resistant, fast-growing, and well-adapted to general environmental conditions, which leads to aquaculture being more sustainable and efficient. This way, genomics will also help identify genes associated with genetic markers that might make some desirable characteristics, like high omega-3 content or superior texture for better seafood products. The genome sequencing projects will identify genetic variation that controls the above traits, and this information will be used to guide breeding programs and lead to improvements in the quality of such products.

Proteomics, or the study of the whole set of proteins produced by the organism, deals with the functional aspect of seafood biology. Proteomics allows for the assessment of variations in seafood proteins due to such factors as diet, stress, or disease. This is critical information describing physiological and biochemical processes that occur in seafood and influence quality and safety. An example of how this can be applied to quality control measures is proteomic analysis for the identification of biomarkers for freshness or spoilage. Similarly, the use of their discovery in nutraceuticals and functional foods is linked with proteomics in the discovery area of novel bioactive peptides and proteins. Such protein profiles are thus useful to unravel compounds bearing certain health benefits like antihypertensive and antioxidant activities.

The study of metabolites, in turn, is covered by the discipline of metabolomics. These are small molecules involved in metabolism. It provides a snapshot of the physiological state of an organism at a particular time. In seafood research, metabolomics is utilized to feature the broad range of metabolites in marine organisms. This technology is very useful in monitoring the influence of various factors on the quality of seafood, for example, the effects of various types of feeds or environmental stressors. Metabolomics is also used in the detection of contaminants and toxins in seafood to ensure its safety. For seafood, metabolite profiling is used to identify quality markers, freshness, as well as possible spoilage, making the process of quality control efficient (Figure 11.2).

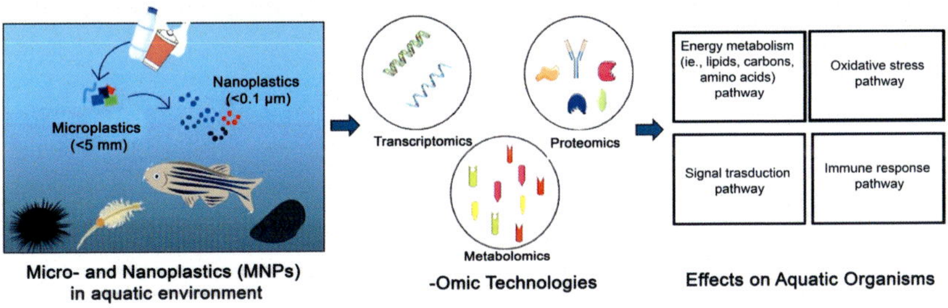

Figure 11.2. *Application of omics approaches for assessing microplastic and nanoplastic toxicity in fish and seafood species.*

Source: Eliso MC, Billè B, Cappello T, Maisano M. Polystyrene Micro- and Nanoplastics (PS MNPs): A Review of Recent Advances in the Use of -Omics in PS MNP Toxicity Studies on Aquatic Organisms. Fishes. 2024; 9(3):98. https://doi.org/10.3390/fishes9030098.

Omics technologies are very strong and have a high future potential in seafood research. High-throughput sequencing, mass spectrometry, and bioinformatics enhancements are continuously boosting the powers of genomics, proteomics, and metabolomics. Together, they can strongly hold a promise to provide full insights into seafood biology, from gene composition to metabolic pathways. In this regard, the integration of genomic data with proteomic and metabolomic profiles will provide an overview of how genetic variation might affect protein expression and the metabolic pathways involved, thus assisting in making informed decisions on breeding, production, and quality control.

Nevertheless, there are also numerous challenges arising from the fact that omics technologies are integrated into seafood research. The first main challenge is brought on by the fact that these technologies are complex and generate significant quantities of data. Omics studies generate a large bulk of information that needs advanced bioinformatic tools and knowledge to analyze and interpret the meaning of the generated data. There is an important task of standardizing methods for the analysis of data derived from omics and associated databases to make sense of this data and bring it to practical use.

These challenges are further added to by the real high cost of omics technologies. The most expensive equipment, reagents, and expertise required in carrying out genomic, proteomic, and metabolomic studies are often what block access to such technologies for smaller research institutions and companies. Hope rests with the current efforts in cost reduction to open up the access and encourage the realization of research benefits in omics.

Moreover, omics technologies often demand interdisciplinary collaborations in seafood research. Such technologies require the junction of expertise in molecular biology, bioinformatics, and seafood science. While allowing for networking in collaborative research projects and collaboration building through the establishment of partnerships with researchers, industry stakeholders, and technology providers, much importance would accrue in addressing this challenge as a way of moving forth research in seafood omics.

The ethical aspect and privacy of genetic information as related to genetic modification also raise ethical issues. The more genomics and other omics technologies are integrated into seafood research and production, the more concerns on ethical and regulatory compliance are important to consider and address, to maintain public trust.

11.5. BIOPRESERVATION TECHNIQUES IN THE SEAFOOD INDUSTRY

Biopreservation techniques are increasingly applied in the seafood industry as a means of extending shelf life, providing safety, and maintaining quality without solely relying on synthetic additives. These techniques represent a more sustainable and potentially healthier alternative, as they exploit natural preservatives and biocontrol agents—which inhibit microbial growth and hence spoilage. Knowledge of the advantages and challenges related to biopreservation is crucial for its effective application in seafood processing.

- **Natural Antimicrobial Agents:** These are preservatives used to extend the shelf life of processed fish and other aquaculture products; they prevent or delay microbial growth and spoilage. Common natural preservatives include essential oils, extracts of plants, and natural acids. For instance, extracts from some herbs like oregano, thyme, and rosemary have been shown to possess essential oils with well-marked antimicrobial effects against pathogenic and spoilage flora. The content of phenolic compounds and terpenes in these oils leads to the disruption of microbial cell membranes and metabolic processes, which in turn enhances seafood preservation.

Plant extracts, especially green tea and garlic, are also used due to their antioxidant and antimicrobial effects. Green tea extract contains polyphenols with free radical scavenging and oxidative stress-reducing activity, which preserves the freshness and quality of seafood. Garlic extract contains a large amount of allicin, which is characterized by strong antimicrobial activity against many forms of spoilage bacteria and fungi, thus helping to assure the safety and prolong the shelf life of the seafood product.

Organic acids, such as citric and lactic acid, reduce the pH of seafood products so they become less supportive of the growth of microorganisms. These are usually accompanied by other methods of preservation to have additional benefits. Lactic acid bacteria, famous for lactic acid production during fermentation, can also be used as natural preservatives. They provide a competitive environment that is disadvantageous to the pathogenic microorganisms. Biocontrol agents refer to the application of microorganisms or their products in controlling spoilage and pathogenic microbes in seafood. They include some good bacteria, bacteriocins, and yeasts competing against or inhibiting micro-organisms that are injurious. In turn, lactic acid bacteria (LAB) have been found to feature in theory an enormous range of applied biocontrol agents due to the possession of the ability to form such antimicrobial substances, among them the ones inhibiting the spoilage bacteria and pathogens. LAB are widely used in fermentation processes, and this application as starter cultures proves to improve the safety and shelf life of the food.

Bacteriocins are antimicrobial peptides produced by some bacteria and appear to have a certain potential for application as biocontrol agents in seafood preservation. For example, nisin is a bacteriocin from the Gram-positive bacterium *Lactococcus lactis*, which has been applied in the inhibition of *Clostridium botulinum*, a pathogen causing botulism. Nisin may be used directly on the food product or combined into packaging materials to provide extended protection from microbial contamination.

Yeasts also serve as biocontrol agents through competitive exclusion in species of Candida and Saccharomyces. In this scenario, this class of yeasts would outcompete and subtract nutrients from spoilage microorganisms, hence making it difficult for their growth, hence increasing the shelf life of seafood. Other yeasts can also form certain antimicrobial compounds that further contribute to biopreservation.

One great benefit to biopreservation is the enhancement of product safety and prolongation of shelf life with fewer synthetic preservatives. One of the significant benefits of applying natural preservatives and biocontrol agents in seafood processing is the reduction of chemical additives, catering to consumer demand for natural, minimal processing of food. The techniques also have the environmental advantage of much fewer resources and less waste compared to classical preservation methods.

Biopreservation can contribute to retaining the quality in terms of both nutrition and sensory seafood products. Normally, natural preservatives and biocontrol agents have relatively lower effects on the sensory characteristics of seafood than some synthetic preservatives, which alter taste, texture, or appearance. Hence, very often, the biopreserved seafood will still exhibit the desirable features, but without the disadvantages of less safety and shelf life.

11.6. CONSUMER PREFERENCES AND MARKET TRENDS IN SEAFOOD

The way people eat and consume seafood has dramatically changed, influenced by changes in dietary patterns, increasing awareness of health benefits, environmental concerns, convenience of preparation, and market trends. The seafood industry is thus responding with product innovation and developing new strategies to cater to the modern consumer, but the factors are changing so fast.

One of the most marked trends that characterize consumer behavior is an increase in preference for healthy and sustainable eating. Consumers have become more aware of health and nutritious values of seafood products since they are rich in omega-3 bodybuilding contents, critical for both heart and cerebral organs. As a result, these fish products have embarked on more marketability sections due to the increased demand and uptake of 'healthy' and 'natural' products. Consequently, the products tend to restore health, like wild-caught fish rich in omega-3, or be low on calories, such as seafood snacks.

Today, with seafood being chosen more and more by consumers, sustainability has become one of the most important aspects to consider. With rapidly growing concerns globally for overfishing and the state of the environment in general, aquaculture is often singled out as a key player in the processes provoking damage to ecosystems, which causes the demand for long-term seafood sourcing to grow rapidly. Customers are increasingly becoming interested in purchasing products certified by councils like the Marine Stewardship Council (MSC) or the Aquaculture Stewardship Council (ASC), which assure them of their responsible harvesting or farming of seafood products. The industry is gradually gearing toward better sustainable practices and more open and transparent means of sourcing products.

The rise in busy lifestyles has set the demand for products that are easy to consume or are semi-processed, in this case, providing time-constrained consumers with convenience. This proliferation of value-added seafood products, including pre-marinated fillets, seafood meal kits, and frozen entrées, especially for quality but conveniently prepared meals, has resulted without forsaking quality. These products are targeted at consumers who wish to derive the health benefits from seafood without the conventional time- and energy-consuming hassles involved in the processing.

Adding to that, innovation in product development, in general, is changing the seafood industry. With consumer tastes constantly changing and demanding novelty even in the ocean, the call of the day is for innovation in products to be novel: feel novel, taste novel, and even novel in format. Therefore, this has given rise to products such as plant-based seafood alternatives, catering to the emergence of consumers who are looking not only for this alternative source of protein but also for a lifestyle change either for personal health reasons or for environmental benefits. This is in addition to new applications of processing techniques and ingredients by seafood companies seeking to supply products into premium or gourmet, adventurous eater consumption categories.

Digital media and e-commerce are also fast-growing trends relevant to seafood consumption. Internet shopping and online transactions have allowed consumers to purchase a selection of seafood products not necessarily found in local markets. This has provided a market for specialized and niche seafood products, for example, exotic types of fish and sustainably farmed shellfish. Convenience in the home delivery system and higher access to in-depth information about the product online are impacting the purchasing decision to a greater extent mainly because consumers are now more knowledgeable about seafood and have become very selective in their purchases.

Demographics and cultural personal characteristics are the current trainers of consumer preferences. All over the world, there is a new appreciation for global cuisines, and this has driven demand for seafood products that play critical roles in these culinary practices. Sushi and sashimi are now common dishes in most of the Western world, and this has translated to a direct call for better high-quality raw fish. The Mediterranean and Asian diets, rich in seafood, have also contributed to increasing the number of seafood consumers in those areas.

The consumption of seafood has increased due to an increase in home cooking, mostly because of the COVID-19 pandemic. Easily cookable seafood at home is generating more demand from consumers. Concerns about food safety and the need for immunity drive consumers to seafood, which is perceived as a safe product providing nutrients.

As a response to these changing consumer preferences, the seafood industry is investing in product innovation and marketing strategies by which to better meet market demands. Companies are developing new products, such as seafood with added nutrients important to a health-conscious consumer or those with low sodium options. They are also looking into sustainable packaging solutions for environmentally conscious buyers. Besides, health benefits, sustainability, and convenience have gained prominence in branding and marketing, respectively, towards purchase-driving factors.

Another trend that is emerging supports traceability and transparency in the supply chain of seafood products. The general consumer is more interested to know where the products have come from, how they were produced, and so on. This has led to the adoption of technologies like blockchain and QR codes to allow consumers to trace the journey of their seafood from ocean to plate. That will enable seafood companies to gain the trust of consumers while providing the needed specifications of their products about origin, sustainability practices, and quality to clearly differentiate themselves in the face of stiff competition.

The growing importance of issues related to sustainability drives the adoption of novel aquaculture practices with a lower environmental burden. For instance, multiple integrated multi-trophic aquaculture (IMTA) systems are developed that can combine the co-farming of fish, shellfish, and even seaweed in a way that is environmentally friendly to its surroundings and produces top-end seafood. Such methods can copy nature and reduce waste and use of resources, consonant with the raised requirements consumers have for sustainable food production.

11.7. GLOBALIZATION AND TRADE IN THE SEAFOOD INDUSTRY

Globalization really hit the seafood industry, and it is the business arena with the most intensive worldwide exchange of food. Seafood products are among the most exchanged food products across the international borderline, and this swift change has made it an international and dynamic business. The benefits from this type of globalization are numerous, from the availability of a wide variety of seafood, more market options, and an ability to fulfill increased global demand. It has also introduced a number of challenges that need to be addressed to ensure that the industry remains sustainable and efficient.

One of the biggest influences that the aspect of globalization has brought about in the seafood industry is that of enhancing market access. Producers from across the globe are nowadays in a position to reach out to consumers in distant markets, a factor that increases sales and revenues. For instance, Norway, China, and Vietnam have established themselves as large seafood exporters, selling a selection of products in Europe, North America, and Asia. With this kind of global scope, seafood producers can take full advantage of demand where domestic production does not meet the consumer requirements.

Globalization has opened the door to high levels of diversity in the seafood people are now able to access. Consumers in one country can enjoy seafood products from across the globe: Norwegian salmon, Thai shrimp, or Japanese tuna. Not only did this bring variety to the consumption of seafood, but it also gave consumers a new dimension of flavor and taste. Without any doubt, globalization played a pivotal role in popularizing seafood as part of world cuisine. Seafood trade across the globe has also offered a platform for developing countries to be linked to the global market. Predominantly, developing countries rely on seafood exportation as a frontline source of foreign exchange and employment. Directly or indirectly, tens of millions of people living within coastal communities depend on the sector for their livelihoods, and that contributes to economic development and poverty reduction in these areas. The way of exporting fishery products helps generate foreign exchange for these nations, very much needed for their economic well-being.

However, globalization and trade in the seafood industry are also ridden with huge problems, with the main one being the environmental impact due to increased seafood production and trade. Demand for seafood has led to overfishing in most regions, depleting fish stock and threatening marine ecosystems. These unsustainable fishing practices, driven by the demand for export to meet international needs, have led to the collapse of certain fish populations and habitat destruction, mainly coral reefs and mangroves. This environmental degradation caused by overfishing not only poses a threat to biodiversity but will also naturally end in the unsustainability of the seafood industry.

The second challenge is the complexity and in many cases opaqueness of seafood supply chains worldwide. It is however rather difficult to trace seafood products, a prerequisite for food safety and labor exploitation, because of the long and fragmented supply chains. In addition, lack of transparency regarding supply chains poses challenges in being able to safeguard that seafood is sourced sustainably and ethically. As more and more consumers become conscientious about the environment and social impacts of their food choices, there is an increasing demand for seafood supply chain traceability and transparency.

Seafood international trade is subject to several regulatory and trade barriers. Numerous country-specific standards and regulations on seafood safety, quality, and sustainability exist, with some countries having low controls and others with stringent procedures, which consequently create barriers to entry for exporters. As an example, in strict-regulation markets by the European Union or the United States, it may be hard for producers from developing countries to access those markets. In addition, trade tariffs, quotas, and other non-tariff barriers may hamper seafood cross-border flows and subsequently affect both the prices and the access of seafood to the market.

In each of the challenges voiced; however, many opportunities are opened up to the seafood industry through globalization. Of these opportunities, market diversification is the most prominent. Further, now that seafood consumers diversifying into new markets apart from one, will help them decrease the risks of market price fluctuations. For example, substantial growth has been realized in eating seafood in the Asian market, particularly in China. The rising middle class in the emerging markets will have an increasing appetite for quality seafood products, hence presenting new growth avenues for consumers.

Advances in logistics and transport technologies also ease the global trade of seafood. These innovations include cold chain logistics, improved packaging, and faster shipping methods, which makeoving longer distances of seafood products possible with minimized loss of freshness and quality. This helps reach global markets for the producers at reduced costs and less environmental damage from transportation. Besides, the growing emphasis on sustainability in the global trade regime leaves opportunities for the seafood industry to inculcate more responsible practices. Sustainable seafood certification, such as that of the MSC or ASC, has gained the status of an important market tool. The main aim behind these certifications is to provide an opportunity to the producer to different produce in the global marketing sphere while demonstrating the handling of produce sustainably. With consumers increasingly becoming greener, there will likely be a growing demand for certified sustainable seafood, providing the industry with an incentive to take action in more sustainable ways.

In line with these, the case of globalization has enabled enhanced international collaboration and sharing experiences among states and organizations within the seafood industry. Interaction – from a joint perspective about fisheries management and aquaculture sustainability issues or exchange, in some cases – is required in every aspect of seafood globalization. The cooperation that may lead to best practices, while sharing innovative technologies, and harmonization of the requirements, can all be expected to give an important input for the advancement of sustainability and increased resilience of the seafood industry at the international level.

11.8. ETHICAL CONSIDERATIONS IN SEAFOOD PRODUCTION

The production of seafood nowadays presents one of the most intensive ethical considerations, with respect to consumers, advocacy groups, and industry stakeholders who increasingly put an emphasis on responsible practices that balance profitability with social and environmental stewardship. These considerations span a wide spectrum across animal welfare, environmental ethics, corporate social responsibility, and consumer expectations, among others.

Animal welfare is a significant ethical issue in the production of seafood, more so within aquaculture, since this has a lot of bearing on the treatment of fish and other aquatic species. Increased awareness comes with increased growth in the industry, thereby ensuring farmed seafood is produced in conditions that minimize stress and suffering. Ethical aquaculture involves sufficient space, clean water, proper feeding, and minimization of antibiotics and chemicals. The main animal welfare concerns in aquaculture production include overcrowding, poor water quality, and inhumane handling during the processes of transport and slaughter. Some producers have even adopted welfare standards and certification schemes concerning the humane treatment of fish. For example, the ASC has stipulations regarding the welfare of fish that include requirements for producers to implement practices that reduce stress and promote the welfare of farmed fish. However, humane treatment in aquaculture is still debated as the science of fish sentience and pain perception evolves:

- **Environmental Ethics:** They are core to questions of ethical seafood production, reflecting concerns around the impact on marine ecosystems. Overfishing, habitat destruction, bycatch, and pollution are significant environmental concerns that occur from both wild capture fisheries and aquaculture. Overfishing reduces the fish stock, disrupting food webs and threatening the long-term sustainability of marine resources. Habitat damage—for instance, the destruction of coral reefs and mangroves by some kinds of fishing gear—amplifies this environmental effect. Bycatch within fisheries is mainly non-selective and is the capture of non-targeted species, which can often be colored by catching endangered animals such as sea turtles and dolphins, and thus might involve ethical concerns in the nature of these fisheries. There are also serious environmental issues regarding pollution from aquaculture operations through nutrient runoff, use of chemicals, and escape of farmed species into the wild. In response, ethical seafood production tries to bring more sustainable methods in fishing, such as quota management, application of selective gear, and ecosystem-based management. For aquaculture, adopting integrated multi-trophic systems—for example, farming many species together, often trying to mimic in structure and function natural ecosystems—might entail less harm to the environment. Certifications like the MSC and ASC promote good environmental stewardship by ensuring the seafood comes from operations that meet stringent sustainability criteria.

- **CSR in Seafood Production:** This implies accountability from the industry towards its stakeholders, workers, the local community, and consumers. CSR in seafood production thus takes care of good labor practices, community engagement, and returns to local economies. The industry has been criticized in cases of labor exploitation, particularly within countries where employees engaged in fishing and processing activities face serious exposure to poor working conditions, low wages, and even cases of forced labor. This means that ethical seafood production calls for companies to enact fair labor standards with safe working environments and workers receiving fair pay. Besides labor rights, CSR also includes engaging with local communities on issues of sustainable development. This could be in the way of local investments, especially in infrastructure development, education, and health services, and technically supporting small-scale fishers and Indigenous communities whose livelihoods depend on fisheries. Companies that perform strongly on CSR issues are also more likely to invest in sustainable practices that are of benefit to the environment and contribute to the long-term viability of the seafood industry.
- **Consumer Expectations:** Regarding the ethics of seafood production have, therefore, shifted to calls for more transparency, sustainability, and social responsibility. At the moment, most consumers relate more to their food sources and the conditions under which it is produced. All these are realizing increasing salience owing to the greater demands they place on the environment and society, coupled with the rising concern for food safety and quality.

Ethical Seafood Production: Consumers are demanding sustainability in the sourcing of products, humane production, and socio-responsibly caught or farmed. For a consumer to believe and be satisfied with how their expectations are being met, there is a requirement that the origin of the catch and meeting of the ethical standards associated with that need to be accessed. This has driven the adoption of traceability systems allowing consumers to be able to track seafood from catch or farm to plate. Certifications and eco-labels, such as MSC,

ASC, and Fair Trade, signal to consumers that a product is meeting certain ethical criteria. In addition, companies are more and more likely to use ethical marketing as a means of differentiation to gain consumers' trust. For example, an appeal to ethically conscious consumers could be made through marketing materials that talked about sustainable sourcing, animal welfare, or fair labor practices.

Bibliography

1. Abuhlega, T. A., & Ali, M. R. (2022). Biogenic amines in fish: Prevention and reduction. *Journal of Food Processing and Preservation, 46*(10). https://doi.org/10.1111/jfpp.16883.
2. Ahmed, N., & Turchini, G. M. (2021). Recirculating aquaculture systems (RAS): Environmental solution and climate change adaptation. *Journal of Cleaner Production, 297*. https://doi.org/10.1016/j.jclepro.2021.126604.
3. Ali, A., Wei, S., Ali, A., Khan, I., Sun, Q., Xia, Q., Wang, Z., Han, Z., Liu, Y., & Liu, S. (2022). Nutritional value of fish. *Encyclopedia.* https://encyclopedia.pub/entry/36388 (accessed on 9 November 2024).
4. Arisekar, U., Shalini, R., Shakila, R. J., Sundhar, S., Banu, A. M. A., Iburahim, S. A., & Umamaheshwari, T. (2024). Trace metals in commercial seafood products (canned, pickled, and smoked): Comparison, exposure, and health risk assessment. *Food Research International, 178*, 113969. https://doi.org/10.1016/j.foodres.2024.113969.
5. Arvanitoyannis, I. S., & Kassaveti, A. (2008). Fish industry waste: Treatments, environmental impacts, current and potential uses. *International Journal of Food Science & Technology, 43*(4), 726–745. https://doi.org/10.1111/j.1365-2621.2006.01513.x.
6. Arvanitoyannis, I. S., & Kotsanopoulos, K. V. (2011). Smoking of fish and seafood: History, methods, and effects on physical, nutritional, and microbiological properties. *Food and Bioprocess Technology, 5*, 831–853. https://doi.org/10.1007/s11947-011-0690-8.
7. Bhagat, J., Zang, L., Nishimura, N., & Shimada, Y. (2022). Application of omics approaches for assessing microplastic and nanoplastic toxicity in fish and seafood species. *TrAC Trends in Analytical Chemistry, 154*, 116674. https://doi.org/10.1016/j.trac.2022.116674.

8. Bhat, Z. F., Morton, J. D., Bekhit, A. E. D., Kumar, S., & Bhat, H. F. (2021). Thermal processing implications on the digestibility of meat, fish, and seafood proteins. *Comprehensive Reviews in Food Science and Food Safety, 20*(5), 4511–4548. https://doi.org/10.1111/1541-4337.12802.
9. Biji, K. B., Ravishankar, C. N., Venkateswarlu, R., Mohan, C. O., & Gopal, T. K. (2016). Biogenic amines in seafood: A review. *Journal of Food Science and Technology, 53*, 2210–2218. https://doi.org/10.1007/s13197-016-2224-x.
10. Boziaris, I. S. (2014). *Seafood Processing: Technology, Quality, and Safety*. Wiley-Blackwell. https://books.google.co.ke/books?hl=en&lr=&id=cqMTAgAAQBAJ&oi=fnd&pg=PR13&dq=Processing+Technologies+for+Seafood&ots=T1fHA5B9UF&sig=W-wAIYWMirMcaUuWZFALyGpGWvA&redir_esc=y#v=onepage&q=Processing%20Technologies%20for%20Seafood&f=false (accessed on 9 November 2024).
11. Calamelli, E., Liotti, L., Beghetti, I., Piccinno, V., Serra, L., & Bottau, P. (2019). Component-resolved diagnosis in food allergies. *Medicina, 55*(8), 490–498. https://doi.org/10.3390/medicina55080498.
12. Campus, M. (2010). High-pressure processing of meat, meat products, and seafood. *Food Engineering Reviews, 2*, 256–273. https://doi.org/10.1007/s12393-010-9028-y.
13. Çankırılıgil, E. C., & Berik, N. (2018). Sensorial evaluation of fish croquettes produced from different seafood. *Aquatic Sciences and Engineering, 33*(3), 96–101. https://doi.org/10.26650/ASE201818.
14. Cardoso, C., Afonso, C., & Bandarra, N. M. (2016). Seafood lipids and cardiovascular health. *Nutrire, 41*(7). https://doi.org/10.1186/s41110-016-0008-8.
15. Çetinkaya, T., Ceylan, Z., Meral, R., Kılıçer, A., & Altay, F. (2021). A novel strategy for Au in food science: Nanoformulation in dielectric, sensory properties, and microbiological quality of fish meat. *Food Bioscience, 41*, 101024. https://doi.org/10.1016/j.fbio.2021.101024.
16. Chen, J., Jayachandran, M., Bai, W., & Xu, B. (2022). A critical review on the health benefits of fish consumption and its bioactive constituents. *Food Chemistry, 369*, 130874. https://doi.org/10.1016/j.foodchem.2021.130874.
17. Cheng, J. H., & Sun, D. W. (2014). Hyperspectral imaging as an effective tool for quality analysis and control of fish and other seafoods: Current research and potential applications. *Trends in Food Science & Technology, 37*(2), 78–91. https://doi.org/10.1016/j.tifs.2014.03.006.
18. Chennu, A., Färber, P., De'ath, G., De Beer, D., & Fabricius, K. E. (2017).

A diver-operated hyperspectral imaging and topographic surveying system for automated mapping of benthic habitats. *Scientific Reports, 7*(1). https://doi.org/10.1038/s41598-017-07337-y.

19. Dong, X., & Raghavan, V. (2022). A comprehensive overview of emerging processing techniques and detection methods for seafood allergens. *Comprehensive Reviews in Food Science and Food Safety, 21*(4), 3540–3557. https://doi.org/10.1111/1541-4337.12987.
20. Erasmus, J., Zimmermann, S., Smit, N., Malherbe, W., Nachev, M., Sures, B., & Wepener, V. (2022). Human health risks associated with consumption of fish contaminated with trace elements from intensive mining activities in a peri-urban region. *The Science of the Total Environment, 825*, 154011. https://doi.org/10.1016/j.scitotenv.2022.154011.
21. Etheridge, S. M. (2010). Paralytic shellfish poisoning: Seafood safety and human health perspectives. *Toxicon, 56*(2), 108–122. https://doi.org/10.1016/j.toxicon.2009.12.013.
22. Fortin, N. D. (2006). The Food Allergen Labeling and Consumer Protection Act: The requirements enacted, challenges presented, and strategies fathomed. Retrieved from: HeinOnline: https://heinonline.org/HOL/LandingPage?handle=hein.journals/jmdl10&div=10&id=&page= (accessed on 9 November 2024).
23. Freitas, J., Vaz-Pires, P., & Câmara, J. S. (2019). Freshness assessment and shelf-life prediction for *Seriola dumerili* from aquaculture based on the quality index method. *Molecules, 24*(19), 3530. https://doi.org/10.3390/molecules24193530.
24. Fu, L., Wang, C., Zhu, Y., & Wang, Y. (2019). Seafood allergy: Occurrence, mechanisms, and measures. *Trends in Food Science & Technology*, 80–92. https://doi.org/10.1016/j.tifs.2019.03.025.
25. Giannetti, A., Pession, A., Bettini, I., Ricci, G., Giannì, G., & Caffarelli, C. (2023). IgE-mediated shellfish allergy in children—A review. *Nutrients, 15*(14), 3112. https://doi.org/10.3390/nu15143112.
26. Gómez, B., Munekata, P. E., Gavahian, M., Francisco, J. B., Bolumar, T., Campagnol, P. C., & Lorenzo, J. M. (2019). Application of pulsed electric fields in meat and fish processing industries: An overview. *Food Research International, 123*, 95–105. https://doi.org/10.1016/j.foodres.2019.04.047.
27. Hachiya, N. (2006). The history and the present of Minamata disease. *The Journal of the Japanese Medical Association, 49*(3), 112–118. Retrieved from: https://www.med.or.jp/english/activities/pdf/jmaj/v49no03.pdf#page=21 (accessed on 9 November 2024).
28. Hassoun, A., Siddiqui, S. A., Smaoui, S., Ucak, İ., Arshad, R. N., Garcia-

Oliveira, P., & Câmara, J. S. (2022). Seafood processing, preservation, and analytical techniques in the age of industry 4.0. *Applied Sciences, 12*(3). https://doi.org/10.3390/app12031703.

29. Hui, Y. H., Cross, N., Nip, W. K., & Stanfield, P. S. (2012). Biochemistry of seafood processing. In B. K. Simpson, *Food Biochemistry and Food Processing* (pp. 344–364). John Wiley & Sons, Inc. https://doi.org/10.1002/9781118308035.
30. Ildiko, O. (2015). Implementation and conversion of an analytical method for the quantification of biogenic amines in seafood products. University of Iceland. Retrieved from: https://hdl.handle.net/1946/20573 (accessed on 9 November 2024).
31. Jaguey-Hernández, Y., Aguilar-Arteaga, K., Ojeda-Ramirez, D., Añorve-Morga, J., González-Olivares, L. G., & Castañeda-Ovando, A. (2021). Biogenic amines levels in food processing: Efforts for their control in foodstuffs. *Food Research International, 144*, 110341. https://doi.org/10.1016/j.foodres.2021.110341.
32. Jairath, G., Singh, P. K., Dabur, R. S., Rani, M., & Chaudhari, M. (2015). Biogenic amines in meat and meat products and its public health significance: A review. *Journal of Food Science and Technology, 52*(11), 6835–6846. https://doi.org/10.1007/s13197-015-1860-x.
33. Jensen, I. J., Mæhre, H. K., Tømmerås, S., Eilertsen, K. E., Olsen, R. L., & Elvevoll, E. O. (2012). Farmed Atlantic salmon (*Salmo salar* L.) is a good source of long-chain omega-3 fatty acids. *Nutrition Bulletin, 37*(1), 25–29. https://doi.org/10.1111/j.1467-3010.2011.01941.x.
34. Kemp, S. E., Ng, M., Hollowood, T., & Hort, J. (2018). Introduction to descriptive analysis. In *Descriptive Analysis in Sensory Evaluation* (Chapter 1). John Wiley & Sons Ltd. https://doi.org/10.1002/9781118991657.
35. Kontominas, M. G., Badeka, A. V., Kosma, I. S., & Nathanailides, C. I. (2021). Recent developments in seafood packaging technologies. *Foods, 10*(5), 940. https://doi.org/10.3390/foods10050940.
36. Kris-Etherton, P. M., Harris, W. S., & Appel, L. J. (2003). Omega-3 fatty acids and cardiovascular disease: New recommendations from the American Heart Association. *Arteriosclerosis, Thrombosis, and Vascular Biology, 23*(2), e1–e6. https://doi.org/10.1161/01.ATV.0000057393.97337.AE.
37. Krittanawong, C., Isath, A., Hahn, J., Wang, Z., Narasimhan, B., Kaplin, S. L., Jneid, H., Virani, S. S., & Tang. W. H. W. (2021). Fish consumption and cardiovascular health: A systematic review. *The American Journal of Medicine, 134*(6), 713–720. https://doi.org/10.1016/j.amjmed.2020.12.017.

38. Kumar, K. P., Kumar, S. P., & Nair, G. A. (2009). Risk assessment of the amnesic shellfish poison, domoic acid, on animals and humans. *Journal of Environmental Biology, 30*(3), 319–325. Retrieved from: https://www.researchgate.net/profile/Achuthan-Nair/publication/41395896_Risk_assessment_of_the_amnesic_shellfish_poison_domoic_acid_on_animals_and_humans/links/0deec51a360280ee61000000/Risk-assessment-of-the-amnesic-shellfish-poison-domoic-acid-on-animals (accessed on 9 November 2024).
39. Li, J. J., Cao, C., & Huang, D. X. (2008). Anti-obesity effects of conjugated linoleic acid, docosahexaenoic acid, and eicosapentaenoic acid. *Molecular Nutrition & Food Research, 52*(6), 631–645. https://doi.org/10.1002/mnfr.200700399.
40. Lund, E. K. (2013). Health benefits of seafood; Is it just the fatty acids? *Food Chemistry, 140*(3), 413–420. https://doi.org/10.1016/j.foodchem.2013.01.034.
41. Mafra, L. L., De Souza, D. A., Menezes, M., Schramm, M. A., & Hoff, R. (2023). Marine biotoxins: Latest advances and challenges toward seafood safety, using Brazil as a case study. *Current Opinion in Food Science, 53*, 101078. https://doi.org/10.1016/j.cofs.2023.101078.
42. Magnussen, O. M., Haugland, A., Hemmingsen, A. K., Johansen, S., & Nordtvedt, T. S. (2008). Advances in superchilling of food—Process characteristics and product quality. *Trends in Food Science & Technology, 19*(8), 418–424. https://doi.org/10.1016/j.tifs.2008.04.005.
43. Malik, G., Arora, R., Chaturvedi, R., & Paul, M. S. (2021). Implementation of genetic engineering and novel omics approaches to enhance bioremediation: A focused review. *Bulletin of Environmental Contamination and Toxicology, 108*(3), 443–450. https://doi.org/10.1007/s00128-021-03218-3.
44. Marques, A., Maulvault, A. L., & Nunes, M. L. (2019). Future challenges in seafood chemical hazards: Research and infrastructure needs. *Trends in Food Science & Technology, 84*, 52–54. https://doi.org/10.1016/j.tifs.2018.09.013.
45. Martínez-Viveros, E. M. G., Dublán-García, O., Saucedo-Vence, K., Morachis-Valdez, A. G., Cira-Chávez, L. A., & Gómez-Oliván, L. M. (2023). Lipid and protein oxidation in *Cyprinus carpio* muscle by environmentally relevant concentrations of glibenclamide. *Water Emerging Contaminants & Nanoplastics, 2*(4). https://doi.org/10.20517/wecn.2023.17.
46. Mathiassen, J. R., Misimi, E., Bondø, M., Veliyulin, E., & Østvik, S. O.

(2011). Trends in application of imaging technologies to inspection of fish and fish products. *Trends in Food Science & Technology, 22*(6), 257–275. https://doi.org/10.1016/j.tifs.2011.03.006.

47. Moerman, F., & Fikiin, K. (2016). Hygienic design of air-blast freezing systems. In *Elsevier eBooks* (pp. 271–316). https://doi.org/10.1016/b978-0-08-100155-4.00020-0.
48. Moreira, L., Milheiro, J., Filipe-Ribeiro, L., Cosme, F., & Nunes, F. M. (2024). Exploring factors influencing the levels of biogenic amines in wine and microbiological strategies for controlling their occurrence in winemaking. *Food Research International, 190*, 114558. https://doi.org/10.1016/j.foodres.2024.114558.
49. Mori, T. A. (2017). Marine omega-3 fatty acids in the prevention of cardiovascular disease. *Fitoterapia, 123*, 51–58. https://doi.org/10.1016/j.fitote.2017.09.015.
50. Mukherjee, S., Horka, P., Zdenkova, K., & Cermakova, E. (2023). Parvalbumin: A major fish allergen and a forensically relevant marker. *Genes, 14*(1), 223. https://doi.org/10.3390/genes14010223.
51. Nag, M., Lahiri, D., Dey, A., Sarkar, T., Pati, S., Joshi, S., Bunawan, H., Mohammed, A., Edinur, H. A., Ghosh, S., & Ray, R. R. (2022). Seafood discards: A potent source of enzymes and biomacromolecules with nutritional and nutraceutical significance. *Frontiers in Nutrition, 9*. https://doi.org/10.3389/fnut.2022.879929.
52. Nagasawa, T., Vora, P., Agwu, N., & Tan, J. (2016). Microfluidic detection of biogenic amines. *Santa Clara University*. Retrieved from: https://scholarcommons.scu.edu/cgi/viewcontent.cgi?article=1047&context=bioe_senior (accessed on 9 November 2024).
53. Naghdi, S., Rezaei, M., Tabarsa, M., & Abdollahi, M. (2023). Ultrasonic-assisted enzymatic extraction of sulfated polysaccharide from skipjack tuna by-products. *Ultrasonics Sonochemistry, 95*. https://doi.org/10.1016/j.ultsonch.2023.106385.
54. Nakazawa, N., & Okazaki, E. (2020). Recent research on factors influencing the quality of frozen seafood. *Fisheries Science, 86*(2), 231–244. https://doi.org/10.1007/s12562-020-01402-8.
55. Narsaria, R., Narsaria, R., & Narsaria, R. (2022). Seafood allergies–Symptoms, diagnosis & treatment. *Medindia*. Retrieved from: https://www.medindia.net/health/diet-and-nutrition/seafood-allergies.htm (accessed on 9 November 2024).
56. Nevado, D. L., Santos, S. D., Bastian, G., Deyta, J., Managuelod, E. J., Fortaleza, J. A., & Jesus, R. D. (2023). Detection, identification,

and inactivation of histamine-forming bacteria in seafood: A mini-review. *Journal of Food Protection, 86*(3). https://doi.org/10.1016/j.jfp.2023.100049.

57. Nguyen, M. V., Arason, S., & Eikevik, T. M. (2013). Drying of fish. *Seafood Processing: Technology, Quality and Safety*. https://doi.org/10.1002/9781118346174.ch7.
58. Nygaard, M. E. (2019). Swedish fermented herring as a marker of rural identity: The Alfta surströmmingsskiva. *Food, Culture & Society, 22*(4), 407–422. https://doi.org/10.1080/15528014.2019.1620585.
59. Ochiai, Y., & Ozawa, H. (2020). Biochemical and physiochemical characteristics of the major muscle proteins from fish and shellfish. *Fisheries Science, 86*, 729–740. Retrieved from: https://link.springer.com/article/10.1007/s12562-020-01444-y (accessed on 9 November 2024).
60. Oehlenschläger, J. (2013). Seafood quality assessment. In *J. Oehlenschläger, Seafood Processing: Technology, Quality and Safety* (Chapter 14). John Wiley & Sons, Ltd. https://doi.org/10.1002/9781118346174.
61. Papadopoulou, C., Economou, E., Zakas, G., Salamoura, C., Dontorou, C., & Apostolou, J. (2007). Microbiological and pathogenic contaminants of seafood in Greece. *Journal of Food Quality, 30*(1), 28-42. https://doi.org/10.1111/j.1745-4557.2007.00104.x.
62. Parlapani, F. F. (2021). Microbial diversity of seafood. *Current Opinion in Food Science, 37*, 45–51. https://doi.org/10.1016/j.cofs.2020.09.005.
63. Power, D. M., Taoukis, P., Houhoula, D., Tsironi, T., & Flemetakis, E. (2023). Integrating omics technologies for improved quality and safety of seafood products. *Aquaculture and Fisheries, 8*(4), 457–462. https://doi.org/10.1016/j.aaf.2022.11.005.
64. Prester, L. (2011). Biogenic amines in fish, fish products and shellfish: A review. *Food Additives & Contaminants: Part A, 28*(11), 1547–1560. https://doi.org/10.1080/19440049.2011.600728.
65. Ralston, N. V. (2009). Selenium health benefit values as seafood safety criteria. *EcoHealth, 5*, 442–455. https://doi.org/10.1007/s10393-008-0202-0.
66. Rathod, N. B., Nirmal, N. P., Pagarkar, A., Özogul, F., & Rocha, J. M. (2022). Antimicrobial impacts of microbial metabolites on the preservation of fish and fishery products: A review with current knowledge. *Microorganisms, 10*(4), 773. https://doi.org/10.3390/microorganisms10040773.
67. Roy, V. C., Islam, M. R., Sadia, S., Yeasmin, M., Park, J. S., Lee, H. J., & Chun, B. S. (2023). Trash to treasure: An up-to-date understanding of the valorization of seafood by-products, targeting the major bioactive

compounds. *Marine Drugs, 21*(9), 485. https://doi.org/10.3390/md21090485.

68. Ruethers, T., Taki, A. C., Johnson, E. B., Nugraha, R., Le, T. T., Kalic, T., & Lopata, A. L. (2018). Seafood allergy: A comprehensive review of fish and shellfish allergens. *Molecular Immunology, 100*, 28–57. https://doi.org/10.1016/j.molimm.2018.04.008.
69. Saravana, P. S., Ummat, V., O'Donnell, C., & Tiwari, B. (2022). Strategies to mitigate protein deficit. In *Food Engineering Innovations Across the Food Supply Chain* (pp. 107–130). https://doi.org/10.1016/B978-0-12-821292-9.00014-5.
70. Schacky, C. V., & Harris, W. S. (2007). Cardiovascular benefits of omega-3 fatty acids. *Cardiovascular Research, 73*(2), 310–315. https://doi.org/10.1016/j.cardiores.2006.08.019.
71. Shahidi, F., Varatharajan, V., Peng, H., & Senadheera, R. (2019). Utilization of marine by-products for the recovery of value-added products. *Journal of Food Bioactives, 6*. https://doi.org/10.31665/JFB.2019.6184.
72. Sharma, A., Sharma, R. K., Kim, Y. K., Lee, H. J., & Tripathi, K. M. (2021). Upgrading of seafood waste as a carbon source: Nano-world outlook. *Journal of Environmental Chemical Engineering, 9*(6), 106656. https://doi.org/10.1016/j.jece.2021.106656.
73. Shibahara, Y., Uesaka, Y., Wang, J., Yamada, S., & Shiomi, K. (2013). A sensitive enzyme-linked immunosorbent assay for the determination of fish protein in processed foods. *Food Chemistry, 136*(2), 675–681. https://doi.org/10.1016/j.foodchem.2012.08.066.
74. Sicherer, S. H., Muñoz-Furlong, A., & Sampson, H. A. (2004). Prevalence of seafood allergy in the United States determined by a random telephone survey. *Food Allergy, Dermatologic Diseases, and Anaphylaxis, 114*(1), 159–165. https://doi.org/10.1016/j.jaci.2004.04.018.
75. Sivamaruthi, B. S., Kesika, P., & Chaiyasut, C. (2020). A narrative review on biogenic amines in fermented fish and meat products. *Journal of Food Science and Technology, 58*(5), 1623–1639. https://doi.org/10.1007/s13197-020-04686-x.
76. Soccol, M. C., & Oetterer, M. (2003). Seafood as functional food. *Engineering and Technology, 46*(3). https://doi.org/10.1590/S1516-89132003000300016.
77. Stammen, K., Gerdes, D., Caporaso, F., & Martin, R. E. (2009). Modified atmosphere packaging of seafood. *Critical Reviews in Food Science and Nutrition, 29*(5), 301–331. https://doi.org/10.1080/10408399009527530.

78. Storelli, M. (2008). Potential human health risks from metals (Hg, Cd, and Pb) and polychlorinated biphenyls (PCBs) via seafood consumption: Estimation of target hazard quotients (THQs) and toxic equivalents (TEQs). *Food and Chemical Toxicology, 46*(8), 2782–2788. https://doi.org/10.1016/j.fct.2008.05.011.

79. Svendsen, E. S., Widell, K. N., Tveit, G. M., Nordtvedt, T. S., Uglem, S., Standal, I., & Greiff, K. (2022). Industrial methods of freezing, thawing and subsequent chilled storage of whitefish. *Journal of Food Engineering, 315*, 110803. https://doi.org/10.1016/j.jfoodeng.2021.110803.

80. Toyofuku, H. (2006). Joint FAO/WHO/IOC activities to provide scientific advice on marine biotoxins (research report). *Marine Pollution Bulletin, 52*(12), 1735–1745. https://doi.org/10.1016/j.marpolbul.2006.07.007.

81. Truswell, A. S. (2010). *Cholesterol and Beyond: The Research on Diet and Coronary Heart Disease 1900-2000*. Springer Science & Business Media. Retrieved from: https://books.google.co.ke/books?id=AZiyeH0LOhEC&dq=r.+Hugh+Sinclair+and+Dr.+John+Dyerberg+spearheaded+the+study+of+these+fatty+acids+in+the+prevention+of+cardiovascular+diseases+during+the+1930s+and+1940s&lr=&source=gbs_navlinks_s (accessed on 9 November 2024).

82. Tsoupras, A., Brummell, C., Kealy, C., Vitkaitis, K., Redfern, S., & Zabetakis, I. (2022). Cardio-protective properties and health benefits of fish lipid bioactives; The effects of thermal processing. *Marine Drugs, 20*(3), 187. https://doi.org/10.3390/md20030187.

83. Turner, A. (2014). Validation of HPLC detection methods for marine toxins. In L. M. Botana (Ed.), *Seafood and Freshwater Toxins* (p. 42). https://doi.org/10.1201/b16662.

84. Umeoguaju, F. U., Akaninwor, J. O., Essien, E. B., Amadi, B. A., Igboekwe, C. O., Ononamadu, C. J., & Ikimi, C. G. (2023). Heavy metals contamination of seafood from the crude oil-impacted Niger Delta Region of Nigeria: A systematic review and meta-analysis. *Toxicology Reports, 11*, 58–82. https://doi.org/10.1016/j.toxrep.2023.06.011.

85. Undeland, I., Lindqvist, H., Chen-Yun, Y., Falch, E., Ramel, A., Cooper, M., & Elvevoll, E. (2009). Seafood and health: What is the full story? In *Marine Functional Food* (pp. 17–87). Wageningen Academic. https://doi.org/10.3920/9789086866588_003.

86. Vasconcelos, H., Coelho, L. C., Matias, A., Saraiva, C., Jorge, P. A., & Almeida, J. M. (2021). Biosensors for biogenic amines: A review. *Biosensors for the Detection of Nitrogen-Based Compounds, 11*(3), 82–100. https://doi.org/10.3390/bios11030082.

87. Venugopal, V. (2021). Valorization of seafood processing discards: Bioconversion and bio-refinery approaches. *Frontiers in Sustainable Food Systems, 5*, 5–29. https://doi.org/10.3389/fsufs.2021.611835.

88. Visciano, P., Schirone, M., Tofalo, R., & Suzzi, G. (2012). Biogenic amines in raw and processed seafood. *Frontiers in Microbiology, 3*. https://doi.org/10.3389/fmicb.2012.00188.

89. Wadhwa, M., & Bakshi, M. (2016). Chapter 10–Application of waste-derived proteins in the animal feed industry. In *Protein Byproducts: Transformation from Environmental Burden into Value-Added Products* (pp. 161–192). https://doi.org/10.1016/B978-0-12-802391-4.00010-0.

90. Wang, F., Zhong, H., & Cheng, J. H. (2022). Comprehensive analysis of the structure and allergenicity changes of seafood allergens induced by non-thermal processing: A review. *Molecules, 27*(18), 5857. https://doi.org/10.3390/molecules27185857.

91. Yan, N., & Chen, X. (2015). Sustainability: Don't waste seafood waste. *Nature, 524*(7564), 155–157. https://doi.org/10.1038/524155a.

92. Yuvaraj, D., Bharathiraja, B., Rithika, J., Dhanasree, S., Ezhilarasi, V., Lavanya, A., & Praveenkumar, R. (2016). Production of biofuels from fish wastes: An overview. *Biofuels for Sustainable Development, 10*(3), 301–307. https://doi.org/10.1080/17597269.2016.1231951.

93. Zarei, M., Maktabi, S., & Ghorbanpour, M. (2012). Prevalence of *Listeria monocytogenes*, *Vibrio parahaemolyticus*, *Staphylococcus aureus*, and *Salmonella* spp. in seafood products using multiplex polymerase chain reaction. *Foodborne Pathogens and Disease, 9*(2). https://doi.org/10.1089/fpd.2011.0989.

INDEX

I

L

M

N

O

P

Q

R

S

T